IMAGES
of America

FIREFIGHTING IN ALLEGANY COUNTY

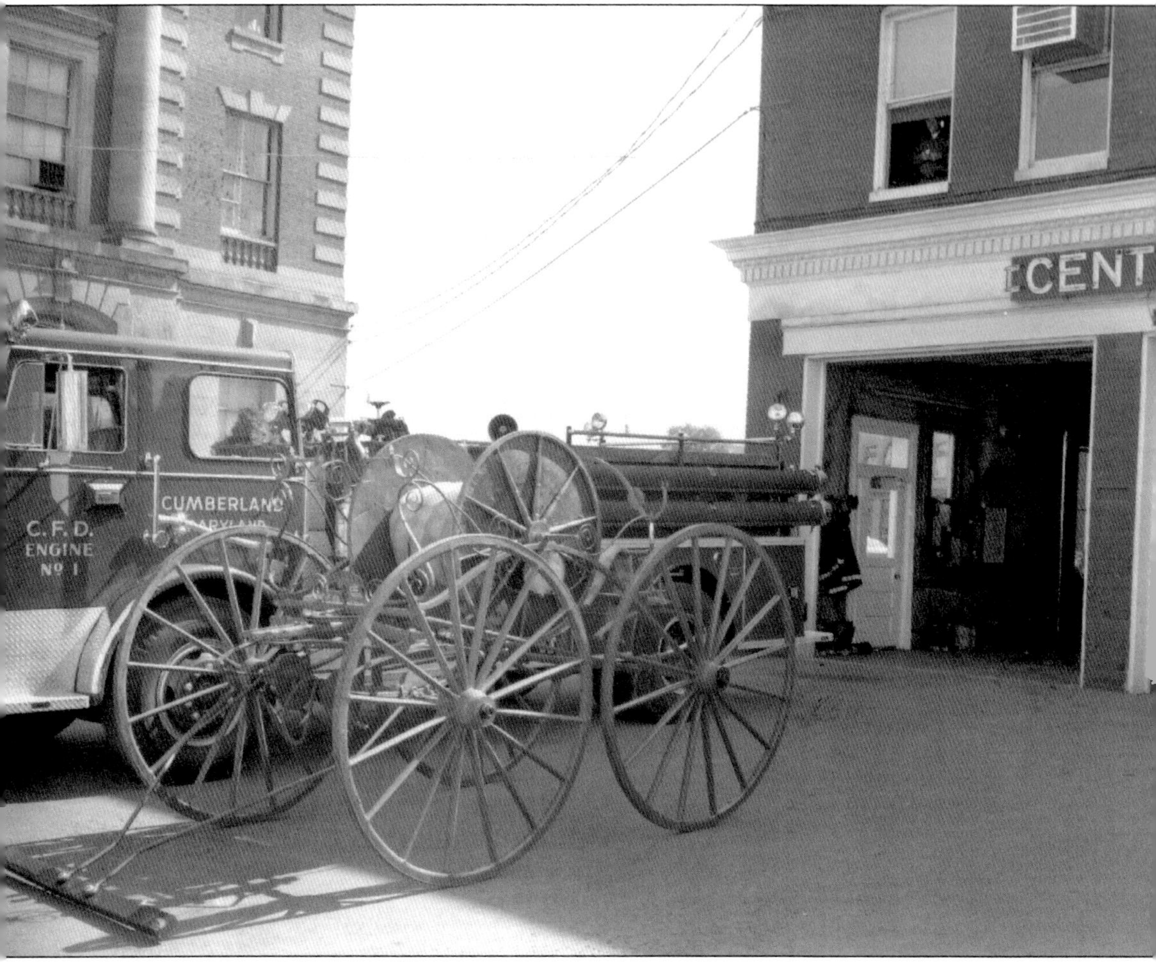

Illustrating the evolution of fire apparatus over 100 years is the 1873 hose reel from Cumberland Hose Company 1 parked beside the 1973 Seagrave pumper. Fire apparatus has only grown larger in the three decades since this photo. (Courtesy Cumberland Fire Department.)

On the Cover: On February 4, 1952, this multi-alarm blaze on Baltimore Street was fought by Cumberland firefighters using the 1949 Seagrave aerial ladder as shown. The fire caused extensive damage to the Potomac Edison Company and Maurice's Department Store.

IMAGES
of America

FIREFIGHTING IN ALLEGANY COUNTY

Warren W. Jenkins

ISBN 0-7385-4197-4

Published by Arcadia Publishing
Charleston SC, Chicago IL, Portsmouth NH, San Francisco CA

Printed in Great Britain

Library of Congress Catalog Card Number: 2005930064

For all general information contact Arcadia Publishing at:
Telephone 843-853-2070
Fax 843-853-0044
E-mail sales@arcadiapublishing.com
For customer service and orders:
Toll-Free 1-888-313-2665

Visit us on the Internet at www.arcadiapublishing.com

This book is dedicated to the following people: Firefighter James Valentine, Cumberland Fire Department, killed September 2, 1907, after the aerial ladder he was climbing broke, dropping him to the pavement. Capt. George W. Miller, Cumberland Fire Department, killed February 22, 1918, when his fire engine was struck by a police patrol wagon. Captain Miller suffered traumatic leg injuries. Firefighter Clyde A. Burley, Corriganville VFC, died December 24, 1991, after suffering injuries the previous November 21 at a fatal traffic accident when he was struck by a vehicle. The author would also like to remember the late Ed Bosanko and Jackson Gerhart—both gentlemen inspired the author's interest in fire service history and enhanced his photo archives.

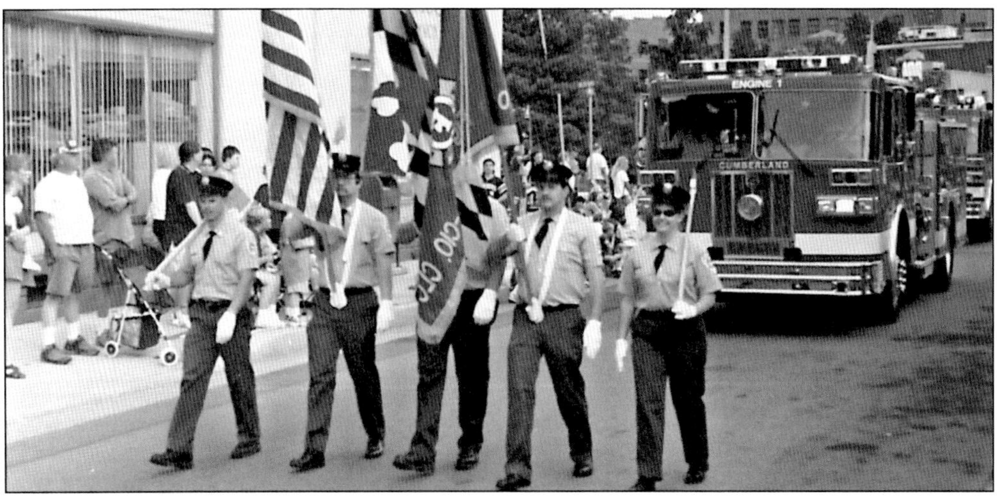

Pictured here is the Cumberland Honor Guard on parade in 2000. (Courtesy Cumberland Fire Department.)

Contents

ACKNOWLEDGMENTS

I would like to thank the following for significant contributions to this book project: Allegany County Museum and curator Joseph Weaver; fire chief William Herbaugh, Cumberland Fire Department; retired fire chief Russell Livengood, Cumberland Fire Department (CFD), and Doris Livengood; and my family for encouragement, support, and photo contributions. Betty Van Newkirk, Frostburg Museum, has been a supporter of my fire department history projects for several years.

On June 18, 2005, a photo shoot was conducted of the CFD at Allegany College. My appreciation to deputy chief Donald Dunn, Lt. Jeff Grogg, and all the members of Chief Dunn's "A" Crew for their cooperation and enthusiasm. On June 25, 2005, Capt. William Davis and Firefighter Bob Burns rendered invaluable assistance on another photo trip. Many thanks to deputy chief Boyce Rogers and administrative assistant Julie Davis for their hospitality and patience as I perused old scrapbooks and albums in their offices at CFD.

A major effort was made over several summer Saturdays to photograph the newest apparatus in Allegany County. I would especially like to thank the following volunteer chiefs and members: Frostburg: Chief Ron Dugan, deputy chief Ken DeVault, and firefighter driver Ken Wilt; LaVale: Chief Steve Ellsworth and firefighters Smith, Long, and Chuck Edmiston; Potomac of Westernport: Chief Bill Kight Jr.; Goodwill of Lonaconing: Pres. James Ternent, assistant chief Woody Nightingale, and members; Oldtown: Chief Ken Wolford; Ellerslie: Chief Dick DeVore and past chief/driver Dick Fetters; Orleans: Pres. Janet Elbin and members; Shaft: Deputy State Fire Marshal Ryan Chapman and members; District 16: Ron Grabenstein Sr. and Ron Grabenstein Jr.; Cumberland firefighters Sam Wilson and Steve Hout for making the Allegany H.I.R.T. truck available for photographs on June 18, 2005; Barton Hose Company and Luke Volunteer Fire Company: Chief Wayne Rounds for past considerations.

Thank you to fellow Arcadia authors Chip Jewell and Justin Mayhue for their support and encouragement.

My appreciation to William Snyder and Joel Woods for their pioneering photographic efforts from 30 to 50 years ago of Allegany County apparatus. On a personal note, many thanks for the support of my colleagues at Sera-Care Bioservices of Frederick, Maryland, particularly Matthew Cummings. Appreciation is also extended to Jeff Mogush and Walter McLaughlin for accompanying me on several trips to Allegany County.

All photos contributed by an individual photographer or an institution are credited to them. All uncredited photos are courtesy of the author or his family.

INTRODUCTION

Allegany County, Maryland, is rich in historical significance. From the French and Indian War in the 1750s to coal mining and transportation, the area has had many interesting books written about this heritage.

This publication will attempt to give an overview of, and pay tribute to, the fire service in Allegany County. The career and volunteer professionals who make up the fire departments in Allegany County have rarely had this type of exposure. Most of the history books about the county or the city of Cumberland have barely mentioned the evolution of fire protection.

There are currently 23 volunteer fire departments and 1 career department in Allegany County. The career fire department in Cumberland was instituted in 1906, and the oldest volunteer company in Frostburg dates from 1878. One volunteer company, Luke, is unique because it protects the town along with one of Allegany County's largest industries, Westvaco (now New Page) paper mill. The company is stationed on paper mill property and is largely staffed by mill employees. There were once several other industrial fire departments at defunct plants such as Celanese, Kelly-Springfield Tires, and Pittsburgh Plate Glass, but little is known about these operations. The Baltimore & Ohio Railroad shops had their own fire brigade for many years.

There are also several independent rescue squads and ambulance services in the county. The EMS service in Frostburg recently became a combination paid-volunteer force. A county-wide hazardous incident team is operated by volunteers from all county departments. The volunteer fireman's association operates an air cascade truck available to all departments and operated by volunteers. The local dispatch center, once a part of local Civil Defense, is now a separate agency.

Fire protection in Allegany County mostly started as citizen bucket brigades that were not organized. They were just trying to keep the fire from spreading beyond the building of origin. After the towns became more developed, inevitably fires would break out in the flimsily constructed wooden structures of the times. Most of the towns in Allegany County had experienced business district fires by the end of the 19th century, and the town of Barton suffered a conflagration in 1919. The people of each town, after suffering these blazes, would organize into volunteer fire companies and try to upgrade their equipment. In most cases, this meant obtaining hand pumpers, hose reels, and ladder wagons. In many of the towns, a water system with fire hydrants was available by 1900. The hose reels would frequently be the only apparatus used with just pressure from the hydrants for hose lines. As late as World War II, a fire company was formed using just a hose reel, and by this time, it was towed by a member's motor vehicle.

Motorization of the various fire companies started in 1916 for Cumberland and in 1917 for Frostburg. The fire companies in the other towns started obtaining motorized apparatus in the 1920s, but most places did not have more than one motor fire truck until after World War II. The first diesel-powered fire apparatus was delivered in the late 1960s. Today the apparatus fleet in Allegany County is probably the most modern ever in service.

Over two dozen apparatus manufacturers have delivered equipment to Allegany County fire departments since the advent of motorized apparatus. This publication will endeavor to present examples of each company's apparatus fleet and the company's quarters. Some photos of major fires have been included, along with other subjects, but a volume several times this size would be needed to present all the big blazes in Allegany County's history, plus the exploits of the firefighters themselves.

This volume is largely the product of the author's archives, but the project was enhanced by access to the Cumberland Fire Department's photo collection. The Cumberland Fire Department celebrates its 100th anniversary in 2006, and it is hoped that the chapter on Cumberland will be of interest.

Due to the assistance from Allegany County Museum, Cumberland, and its curator, Joseph Weaver, the author urges any individual, family, or fire company that would like to enhance the preservation of fire service history to consider contacting him to contribute items to the museum.

While most fire service terms in this book have been simplified, the acronym GPM refers to gallons per minute, the rated capacity of the pump on a fire pumper. For example, the pump can deliver 500 gallons per minute, usually necessary only when in action at a fire or during testing.

One

Queen City Firefighting

Cumberland, Maryland, was once known as the Queen City of the Alleghenies. Although Cumberland has lost approximately half of its population and much of its industrial base, the Cumberland Fire Department has become and remains a modern and well-equipped emergency agency.

Cumberland was protected by at least six volunteer companies prior to the paid fire department's inception on March 26, 1906. The Pioneer Hose Company used the only steam engine owned by a Cumberland fire company, but their firehouse was razed to build the Central Fire Department No. 1 in 1905–1906. The Cumberland Hose Company was located on North Mechanic Street, and their building is still in existence as is the Chapel Hill Hose Company house on Arch Street in South Cumberland. Three other companies, South Cumberland Hose Company on Thomas Street, the Vigilant Hose Company on Walsh Alley and Washington Street, and Friendship Hose Company on Broadway have vanished without much trace. Although the volunteer companies in Cumberland were rapidly phased out of firefighting duties, two companies (Cumberland Hose and Chapel Hill) remained chartered organizations with state recognition at least through 1947.

Cumberland suffered several conflagrations prior to the start of the paid fire department in 1906, with the last occurring in 1893. The idea of a paid fire department first appeared with community meetings in August 1902. Over the next four years, considerable effort went into constructing buildings, hiring men, and buying equipment. Three American LaFrance chemical and hose carts were purchased along with a 65-foot horse-drawn Seagrave wooden aerial ladder.

The original two buildings of the Cumberland Fire Department (CFD) were Central No. 1 on the site of Pioneer Hose and South End No. 2 on Browning Street. The original West Side No. 3 station on Spruce Place was built in 1907. In the mid-1920s, three bungalow-style fire stations were built to replace stations No. 2 and No. 3 and to open new Engine No. 4 on Frederick Street north of the B&O Railroad. In 1978, Central No. 1 was razed following construction of the new Public Safety Headquarters. The following year, West Side No. 3 was closed, and then in 2003, East Side station was closed. Future station construction may be at a location east of the city to cover recently annexed city property.

The Cumberland Fire Department has enjoyed stable leadership over the past 50 years with just four men holding the position of chief over that time.

As the CFD approaches its 100th anniversary as a paid department in 2006, the department is striving to continue the tradition of superb fire protection along with skilled emergency medical services. The department's 65 members are involved in Haz-Mat teams, high angle rescue, and swift water rescue, and they are enthusiastic "interior" firefighters. The citizens of Cumberland are fortunate to have a professional organization as the CFD to protect them from disaster or medical emergency.

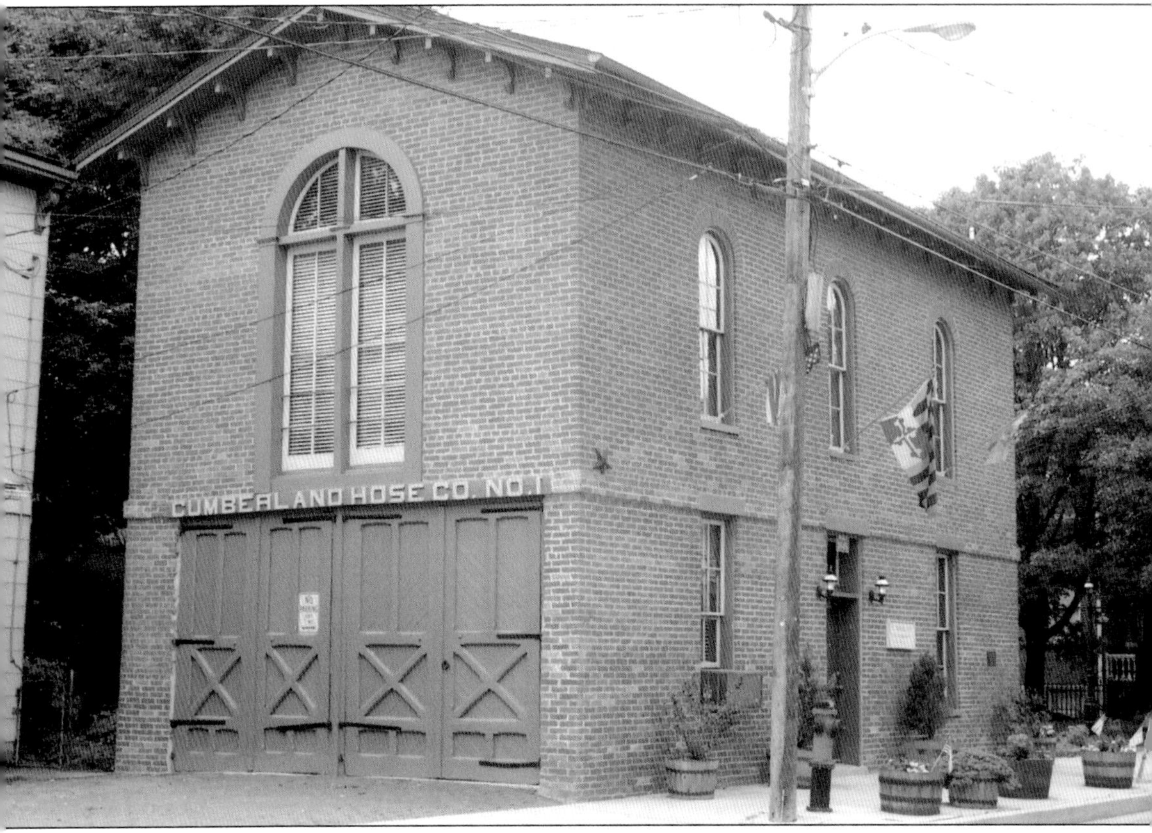

Here is Cumberland Hose Company No. 1's firehouse in 2005. The building has been restored and remodeled inside as office space.

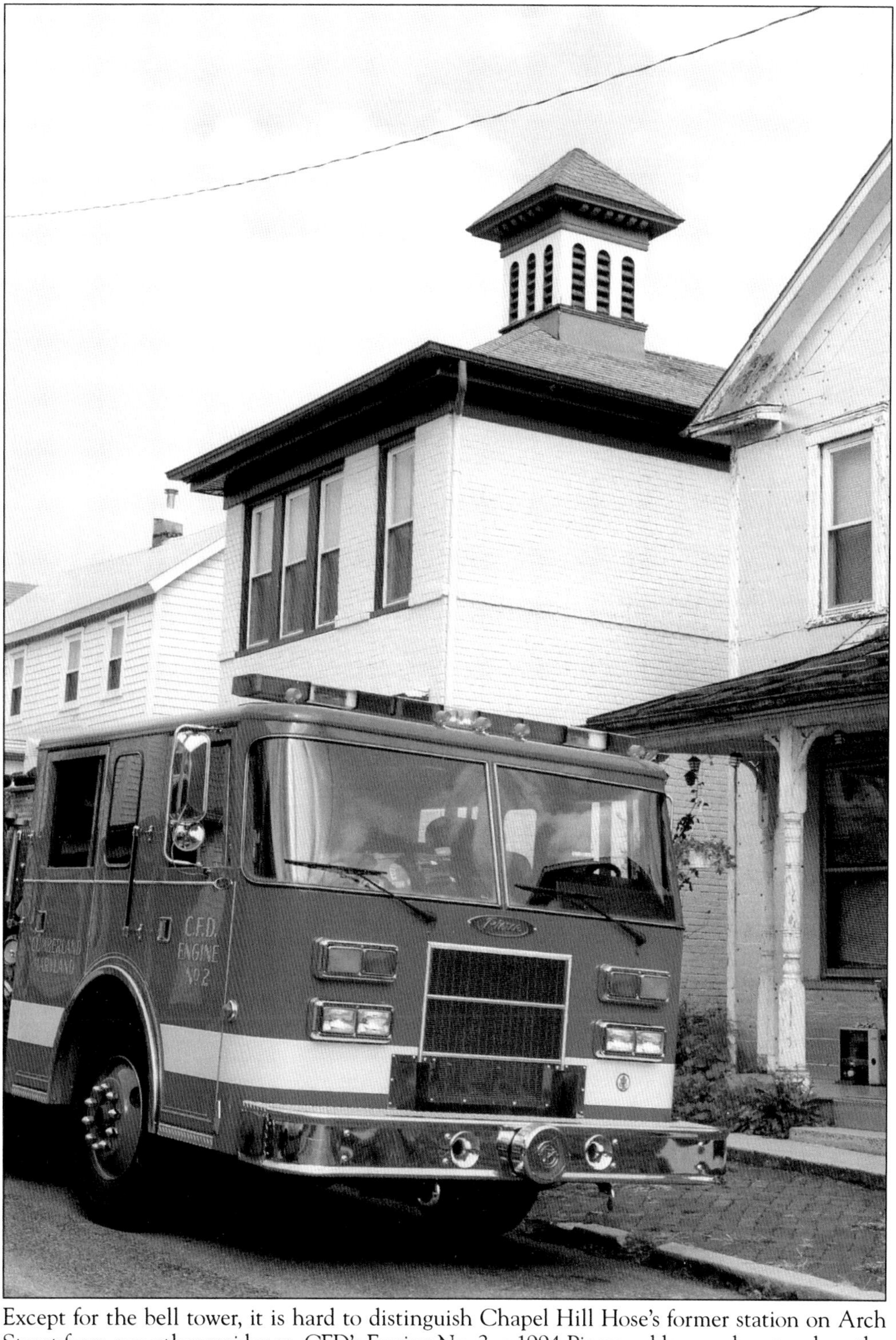

Except for the bell tower, it is hard to distinguish Chapel Hill Hose's former station on Arch Street from any other residence. CFD's Engine No. 2, a 1994 Pierce, adds a modern touch as the firefighters distribute smoke alarms in 2003.

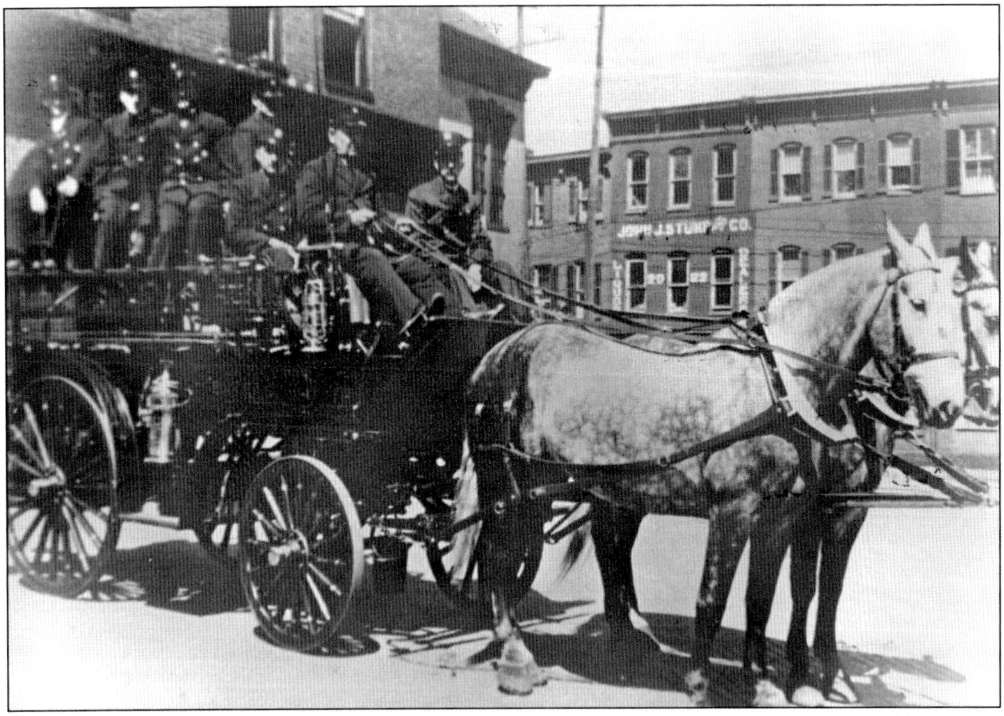

Loaded with most of the active and paid firefighters in 1906, this was the chemical and hose cart used at Central Station until 1916. (Courtesy Cumberland Fire Department.)

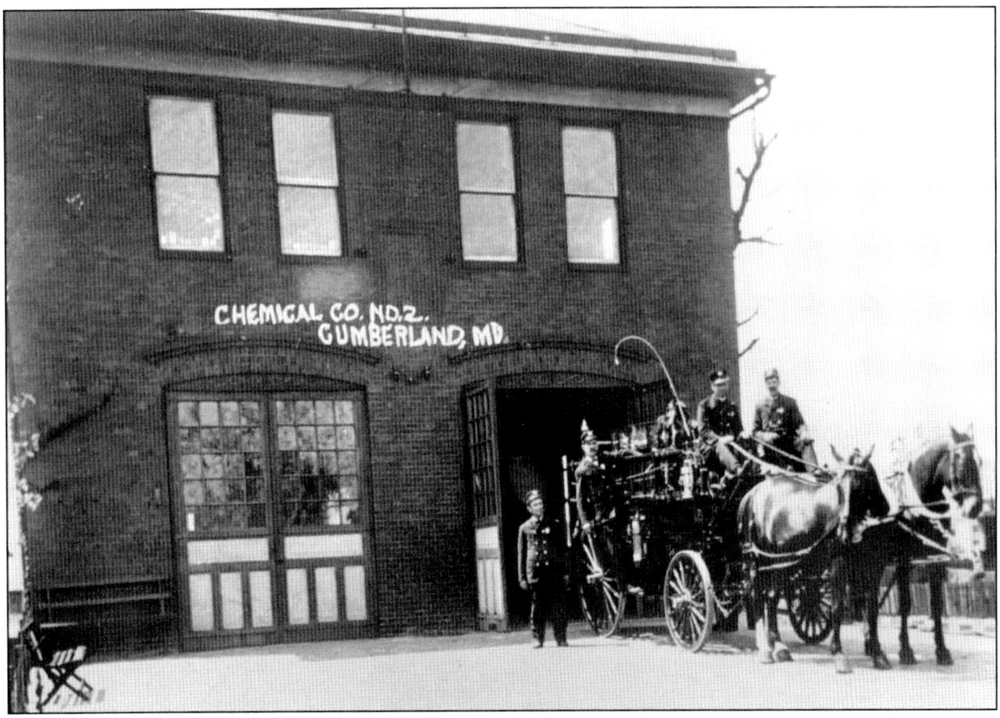

This station on Browning Street in South Cumberland operated from 1906 to 1927. It still exists but has been modified beyond recognition as a fire house. (Courtesy Cumberland Fire Department.)

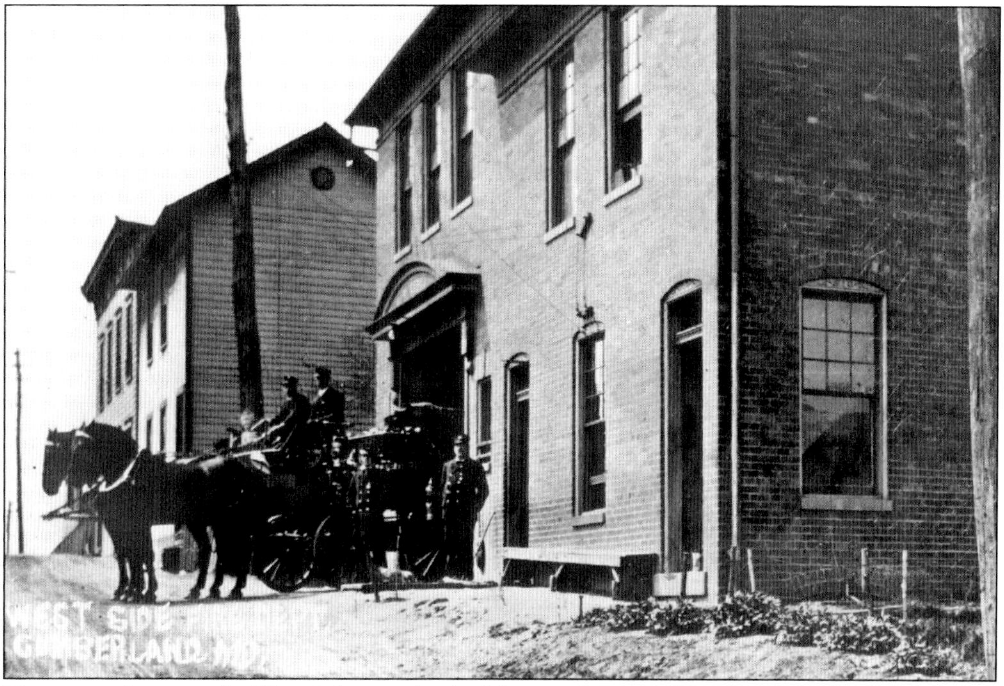

The original station for West Side was this building on Spruce Alley. It operated until 1927, and the station was converted to a residence, now almost unrecognizable. (Courtesy Cumberland Fire Department.)

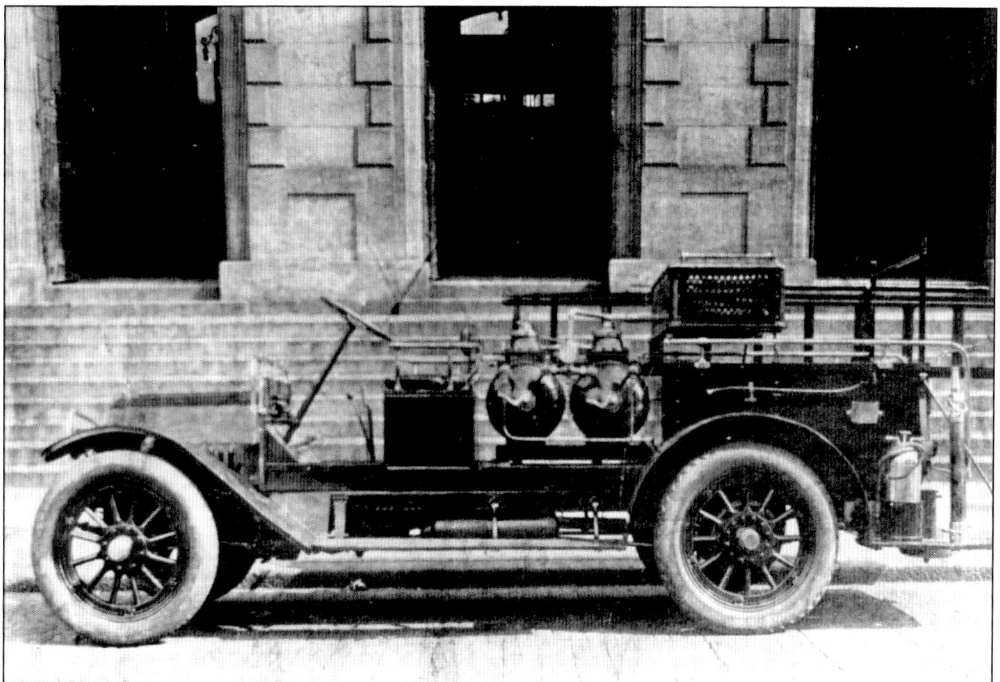

This 1916 Buick chassis combined with the 1906 chemical cart was the original motorized apparatus used by the CFD at Central Station. (Courtesy Cumberland Fire Department.)

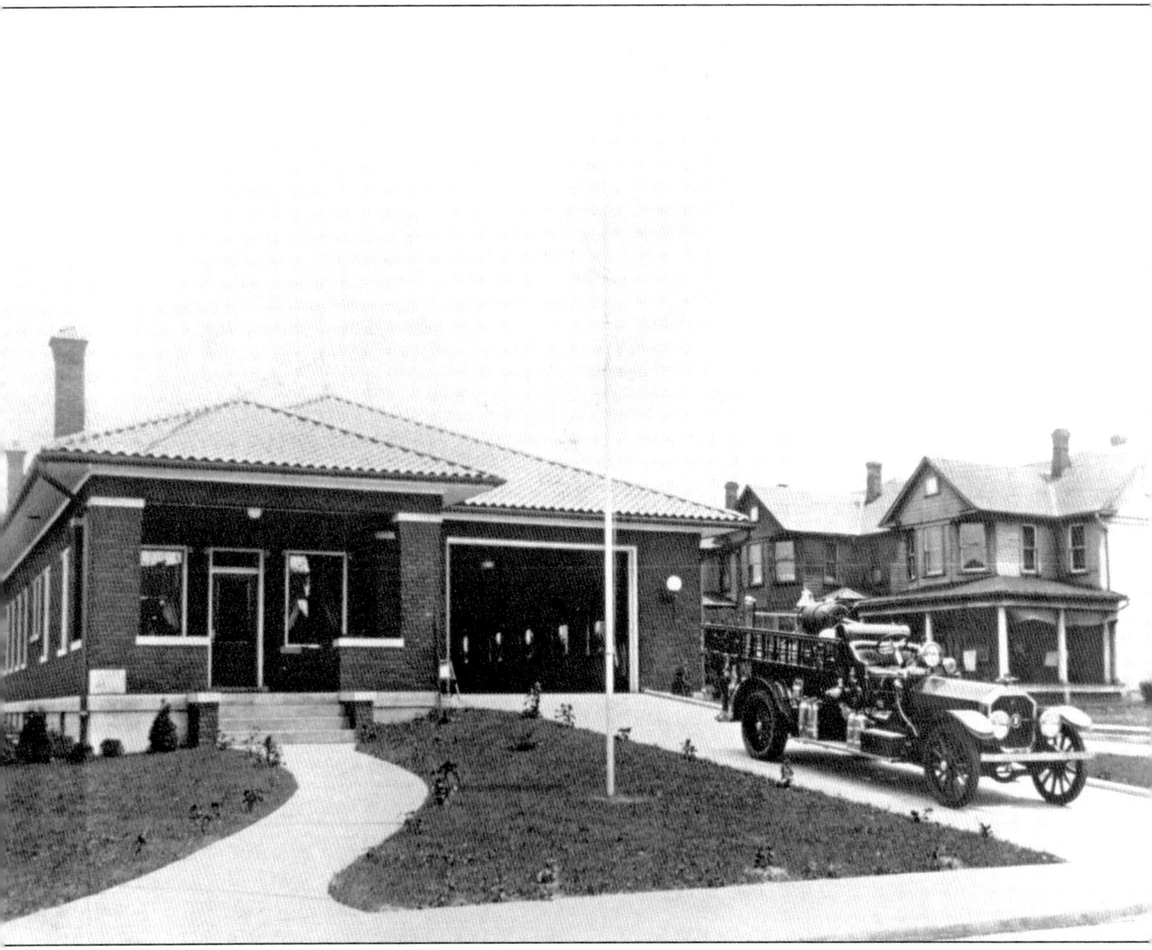

The bungalow-style station for South Cumberland was opened in early 1927. The American LaFrance pumper shown was probably the 1917 model that was the first motorized pumper for CFD. (Courtesy Cumberland Fire Department.)

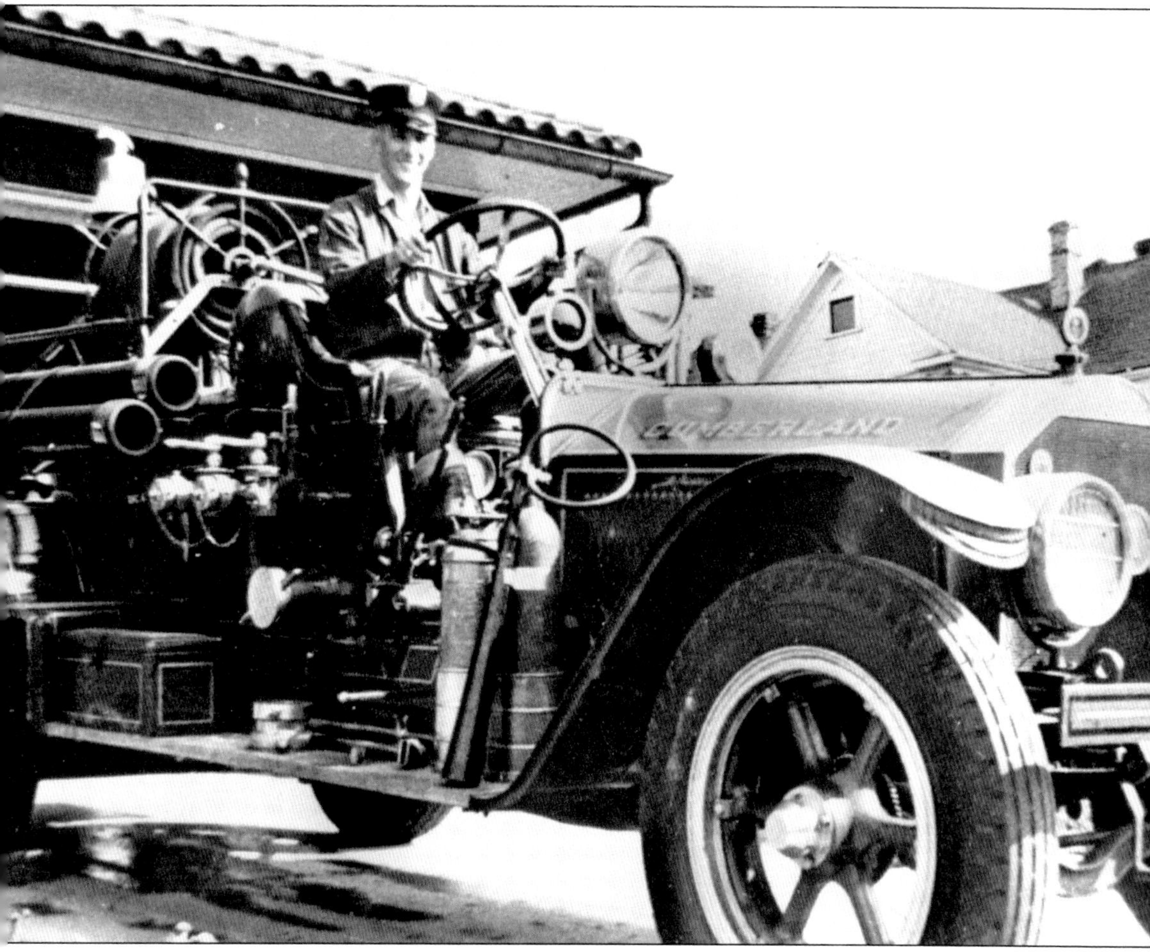

Longtime CFD fireman Robert Gormer is pictured driving one of CFD's early American LaFrance pumpers. CFD purchased five pumpers from American LaFrance from 1917 to 1934. (Courtesy Cumberland Fire Department.)

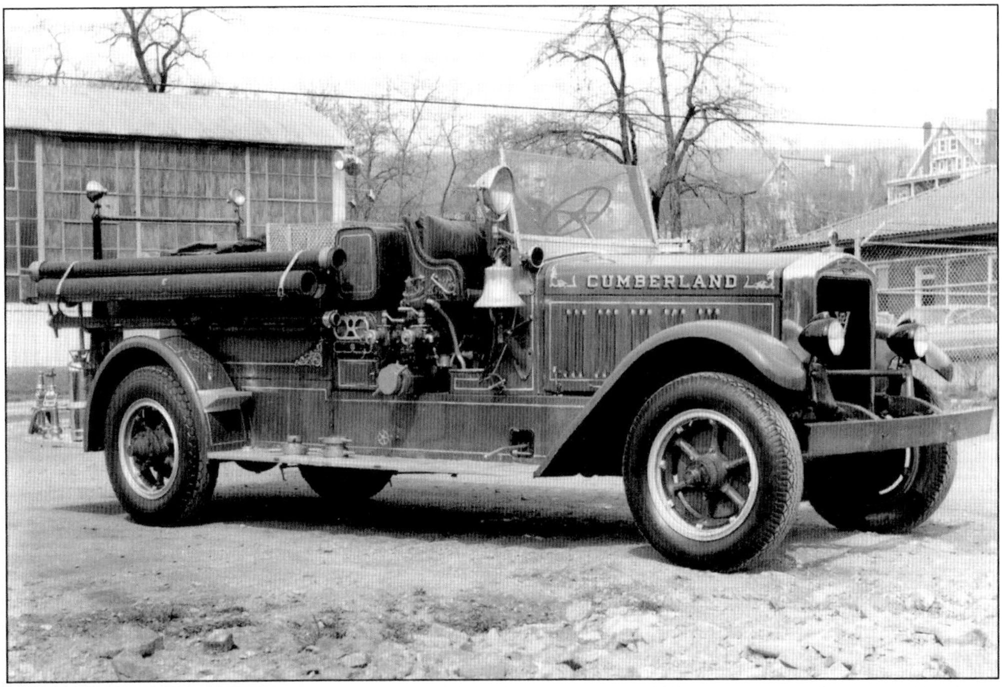

Here is CFD's 1934 American LaFrance 750-GPM pumper in 1973, almost at the end of its long career. The driver is George Hoblitzell, who was the apparatus driver at West Side for many years. (Courtesy Richard Adelman.)

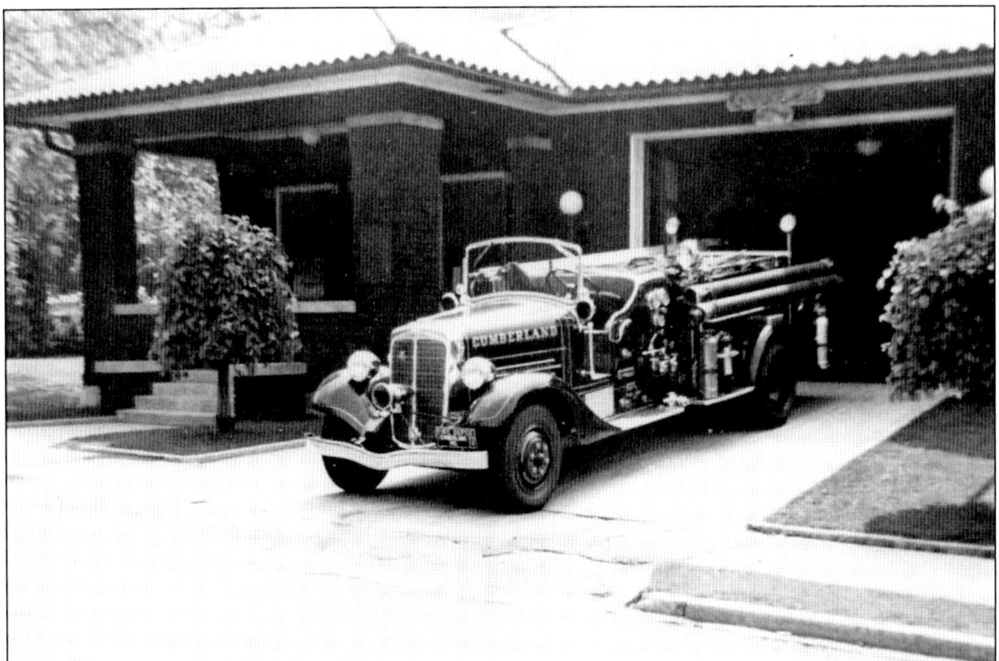

Shown is CFD's 1937 Ahrens-Fox, which had a 600-GPM pump and served as the East Side Station's pumper from 1937 until 1962. It was a WC model (Reg. 9018). (Courtesy Cumberland Fire Department.)

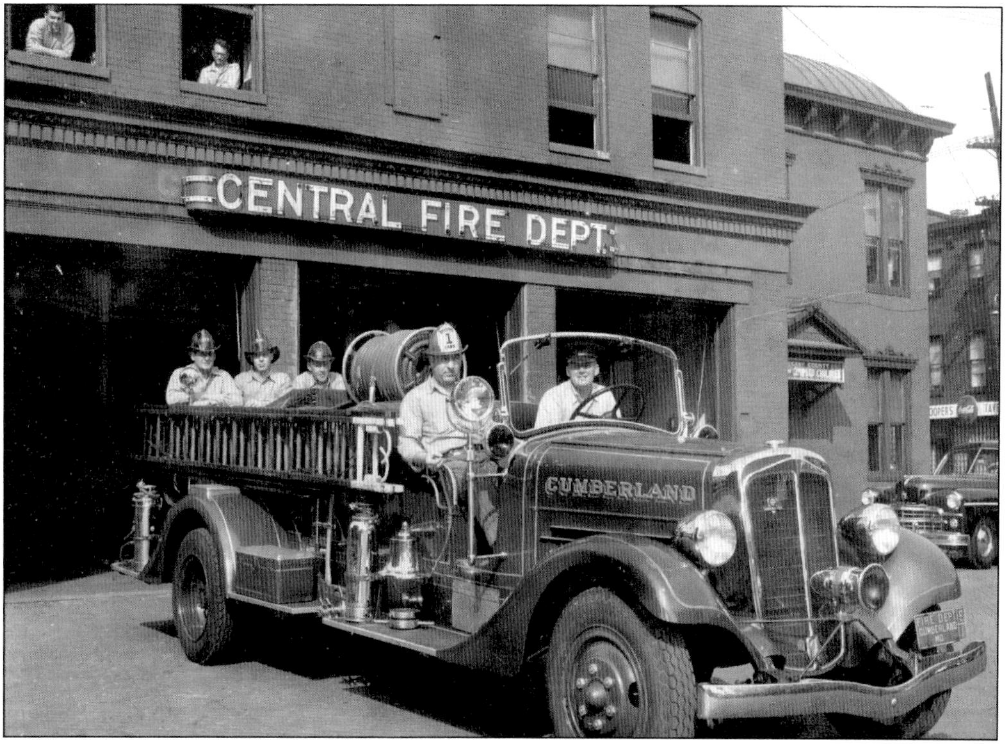

Central Station received this 1938 Ahrens-Fox VC model (Reg. 9041), which had a 600-GPM pump. Pictured in the cab are Capt. Vincent Wegman (left) and driver John Morrissey. Pictured from left to right on the tailboard are H. Cunningham, R. Martin, and W. Hartung. (Courtesy Cumberland Fire Department.)

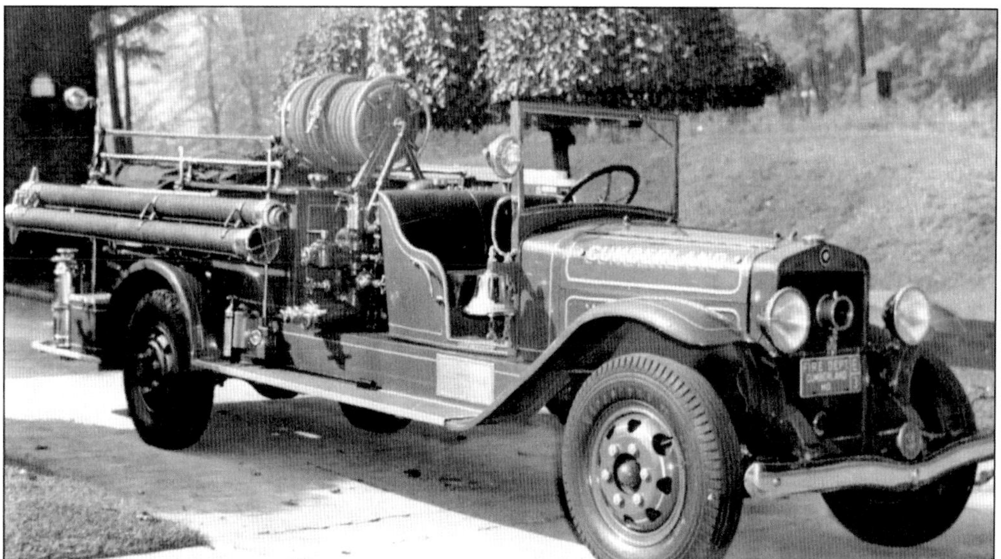

In December 1942, Ahrens-Fox delivered this rebuilt 1931 Model V pumper (Reg. 7021), which had a 500-GPM pump. After serving at West Side Station for a few years, it was eventually placed at Constitution Park. (Courtesy Cumberland Fire Department.)

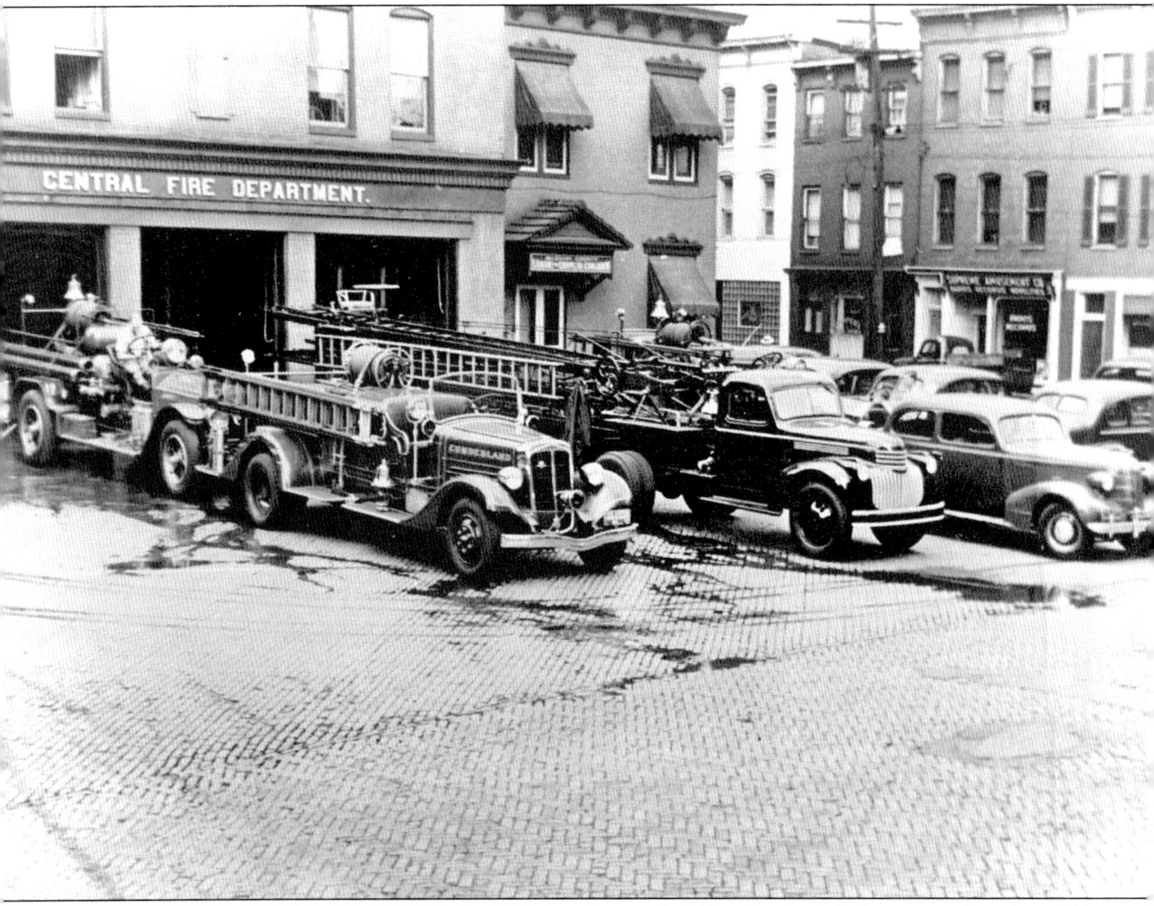

This is the Central Station apparatus fleet in the mid-1940s. In the left foreground is the 1938 Ahrens-Fox; behind is a reserve American LaFrance; in the center is a 1945 Chevrolet tractor with the 1907 Seagrave aerial ladder. On the right is a Pontiac fire chief's car with another reserve American LaFrance behind. (Courtesy Cumberland Fire Department.)

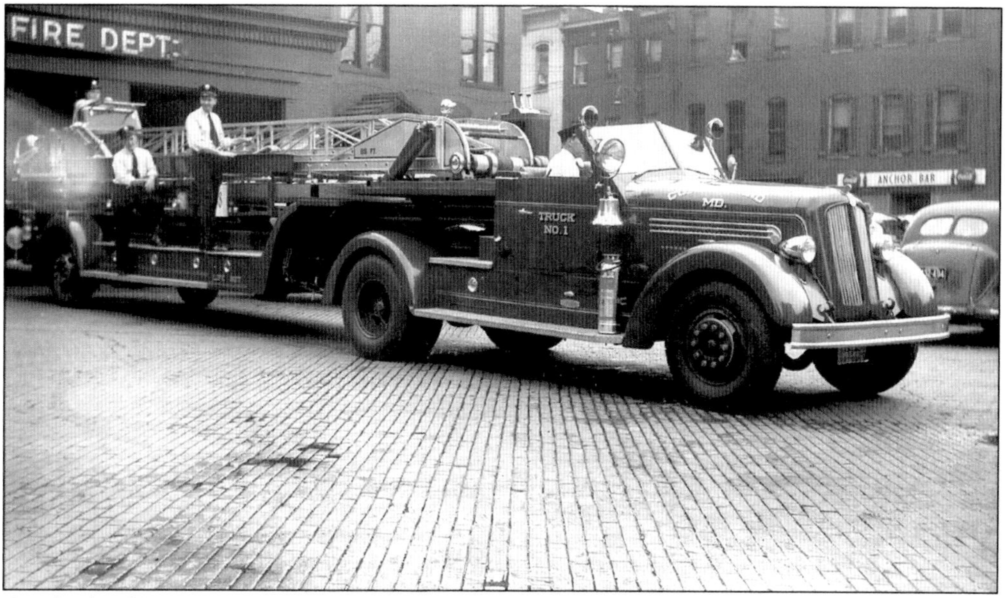

Cumberland Fire Department received a modern steel aerial ladder truck in 1949 when Seagrave delivered this big 85-foot aerial. (Courtesy William Snyder.)

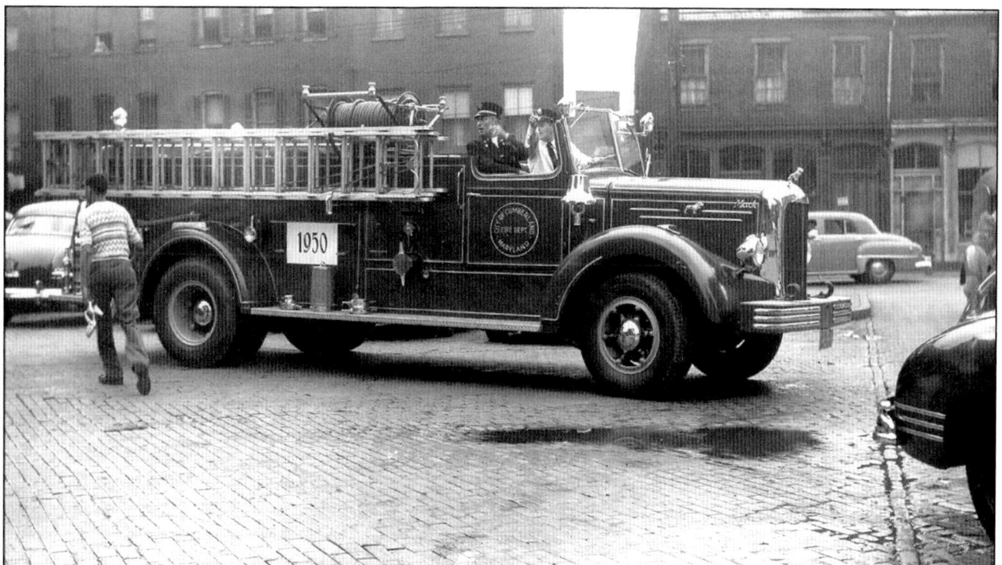

CFD obtained a new 1950 Mack L model pumper for the West Side Station. It could pump 750 GPM and carry 150 gallons of water. It is still on the department's roster and occasionally displayed. (Courtesy William Snyder.)

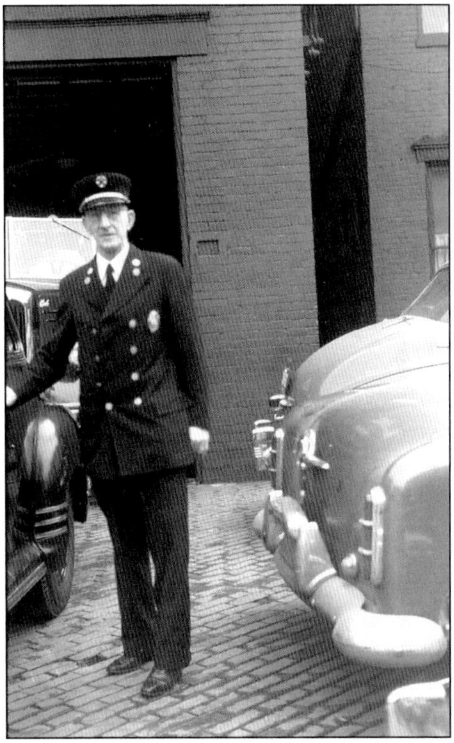

Jeremiah T. Defibaugh, a veteran member of CFD, served as chief for only about two months in 1951 before he retired. After the death of CFD's chief, John E. Brennenin, in December 1950, three senior ranking officers served as chief over the next nine months—William Lindner (March 1, 1951–April 1, 1951), Paul Reid (April 1, 1951–May 14, 1951), and Defibaugh (May 14, 1951–July 20, 1951)—allowing each to retire with the maximum benefits for the position. A new chief, Virgil Parker, was finally named in July 1951, and he served until 1967. (Courtesy William Snyder.)

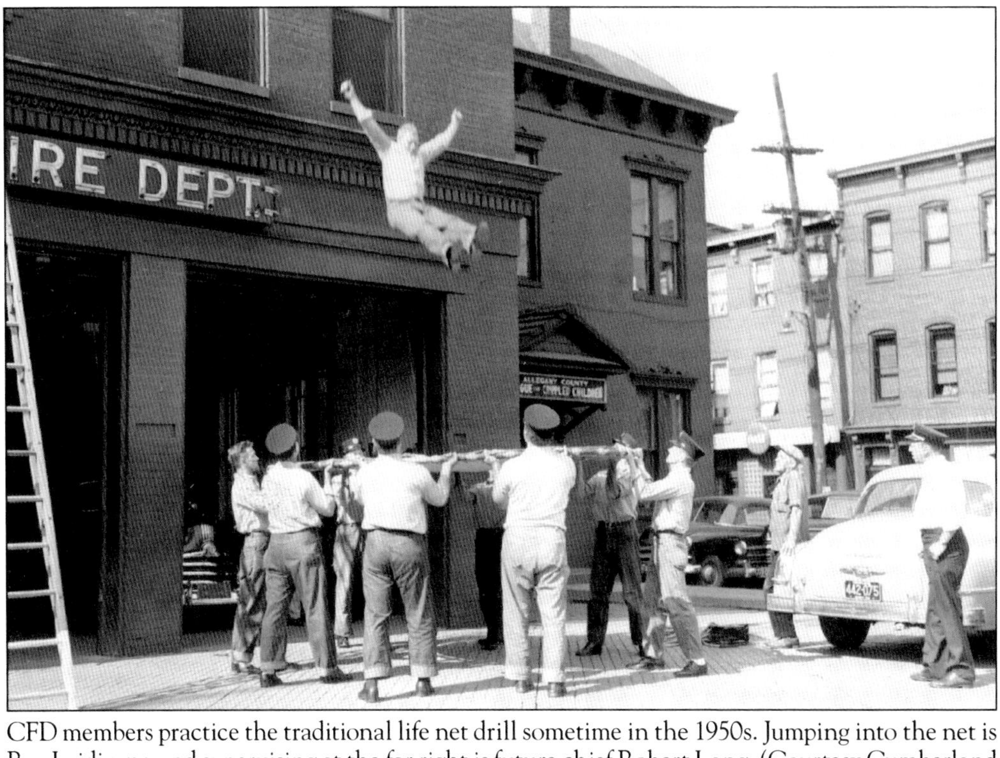

CFD members practice the traditional life net drill sometime in the 1950s. Jumping into the net is Ray Leidinger, and supervising at the far right is future chief Robert Long. (Courtesy Cumberland Fire Department.)

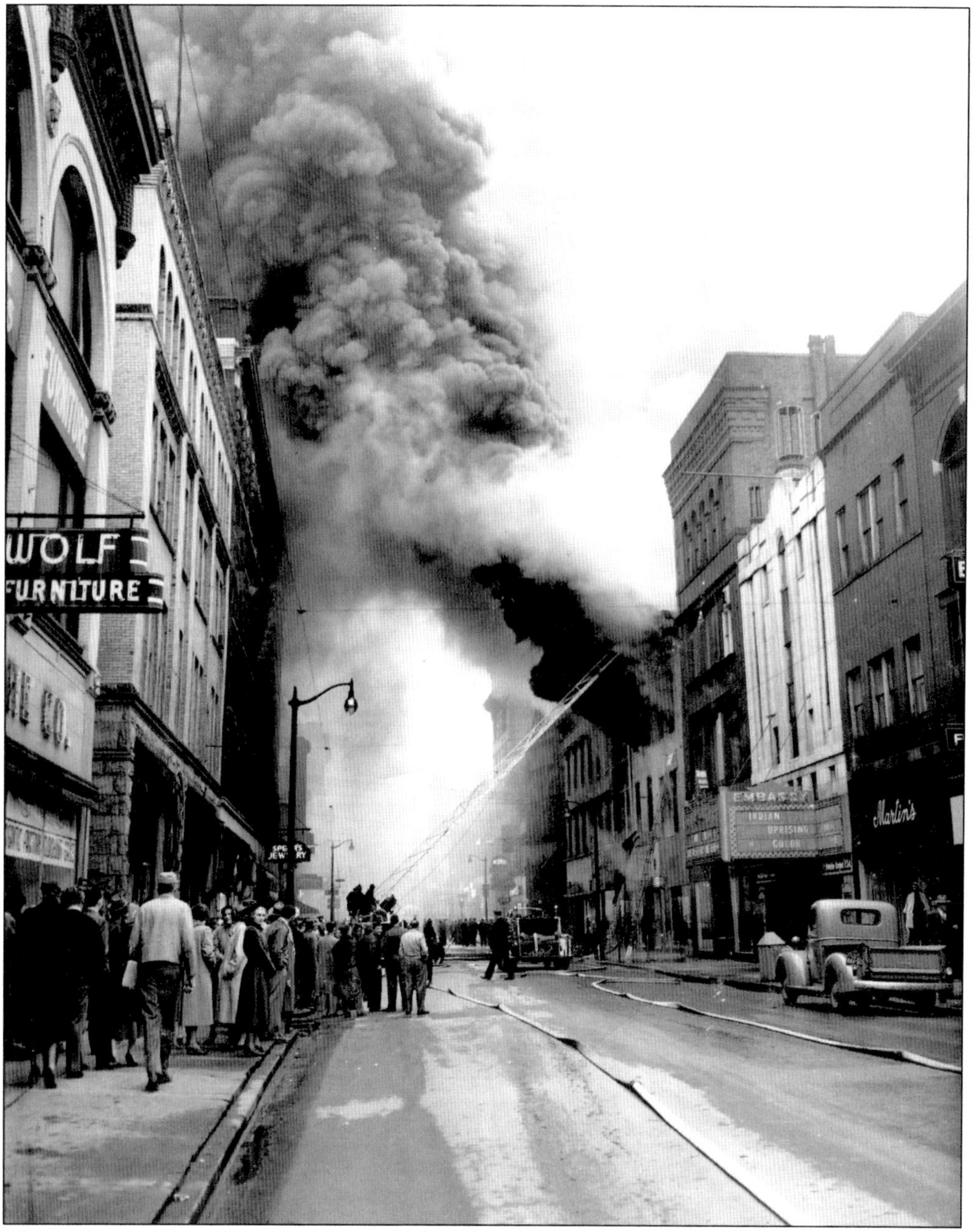

This is the opposite side of the photo used on the cover of this book. It shows the February 1952 Baltimore Street fire. The blaze was fought by CFD and several nearby volunteer companies without radios and modern breathing apparatus. (Courtesy Cumberland Fire Department.)

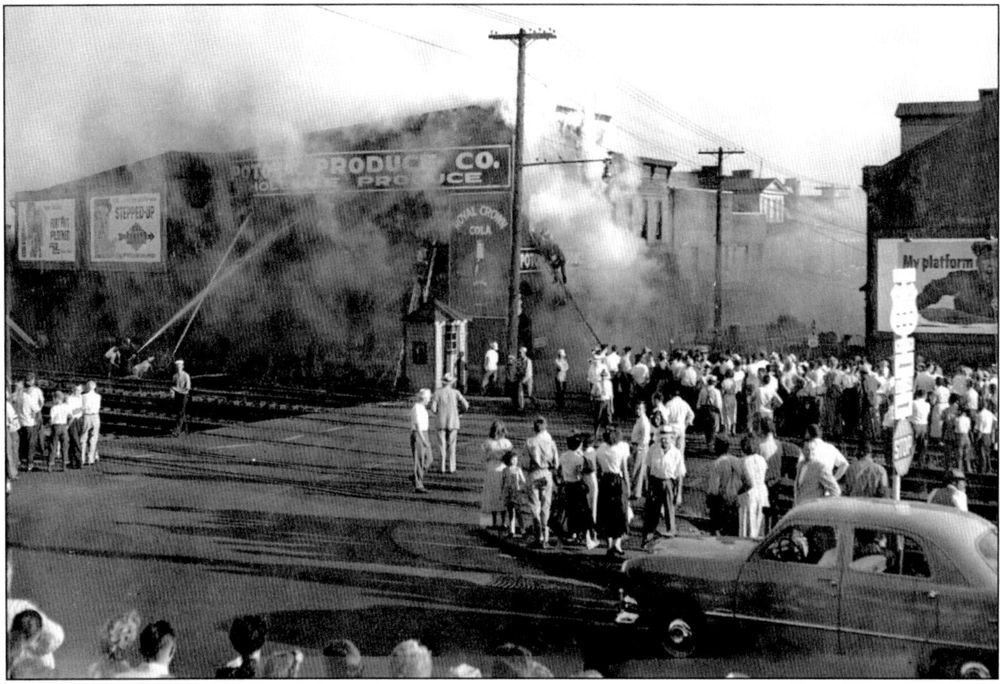

On a very hot July 24, 1952, the Potomac Produce warehouse burned on North Bedford Street at the B&O railroad crossing. This is where the present-day McMullen Bridge is located. This blaze injured several firemen. (Courtesy Cumberland Fire Department.)

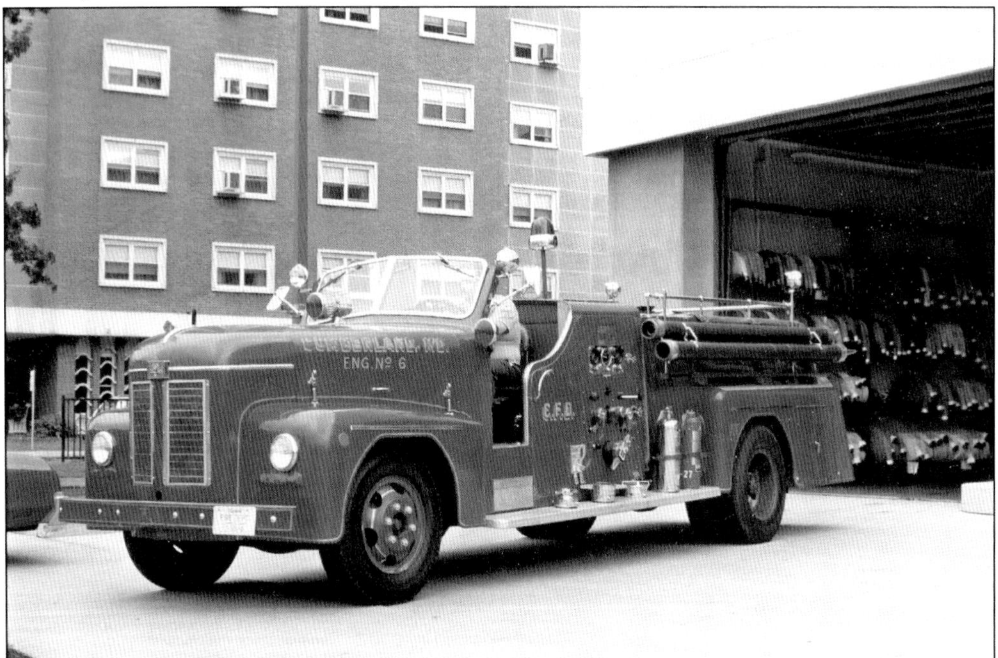

In 1954, CFD received this Ahrens-Fox EC model (Reg. 9165), which could pump 750 GPM and carry 150 gallons of water. It was sold to a collector in Morgantown, West Virginia, in 1989. (Courtesy John Floyd.)

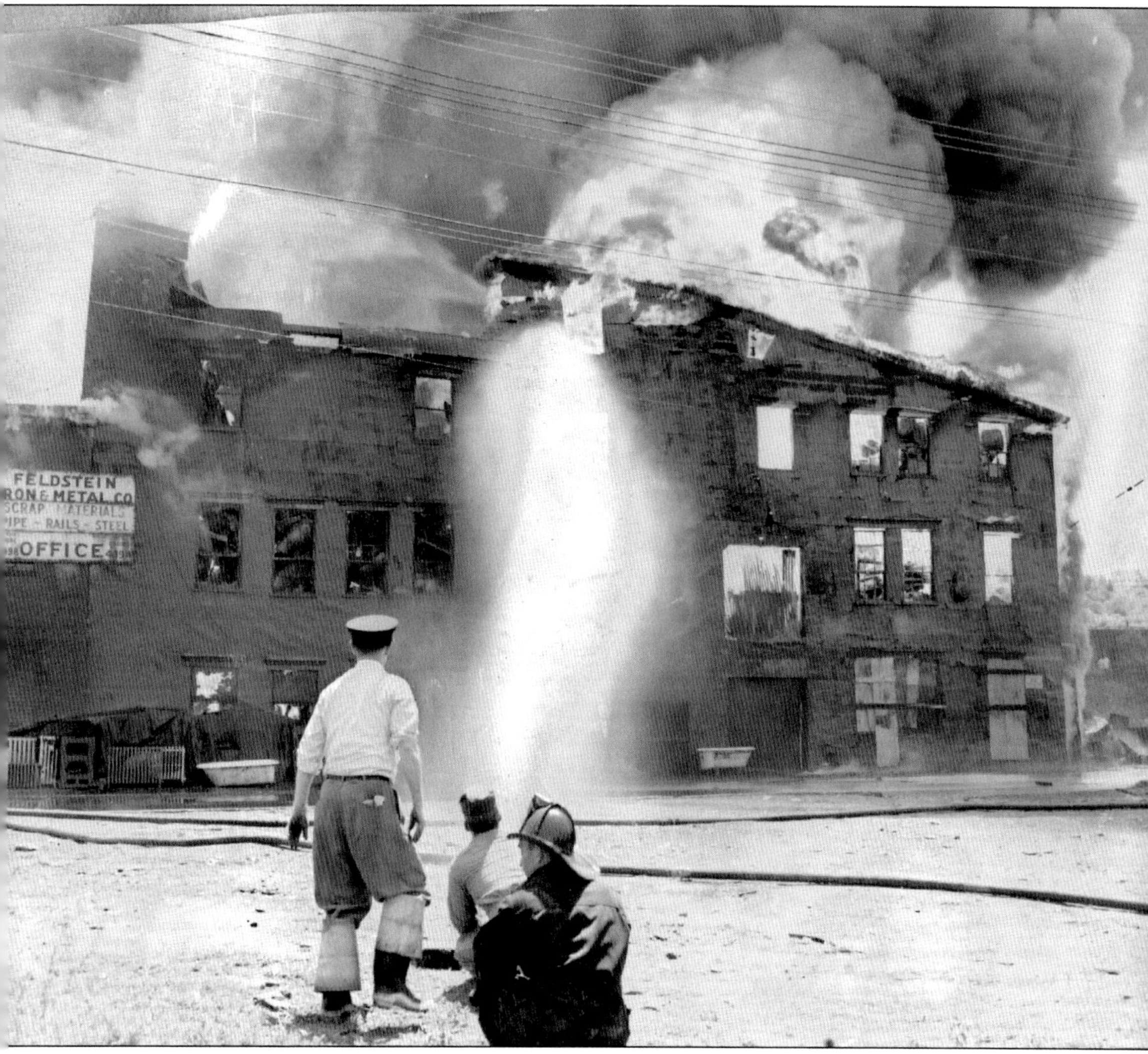

On June 4, 1955, Feldstein's Scrap Yard on Wineow Street burned in a spectacular fashion. CFD members fought the blaze from a distance because of various hazards on the property, including a flimsy tin-and-wooden structure, junked automobiles, unknown hazardous chemicals in old appliances, and gas tanks. (Courtesy Cumberland Fire Department.)

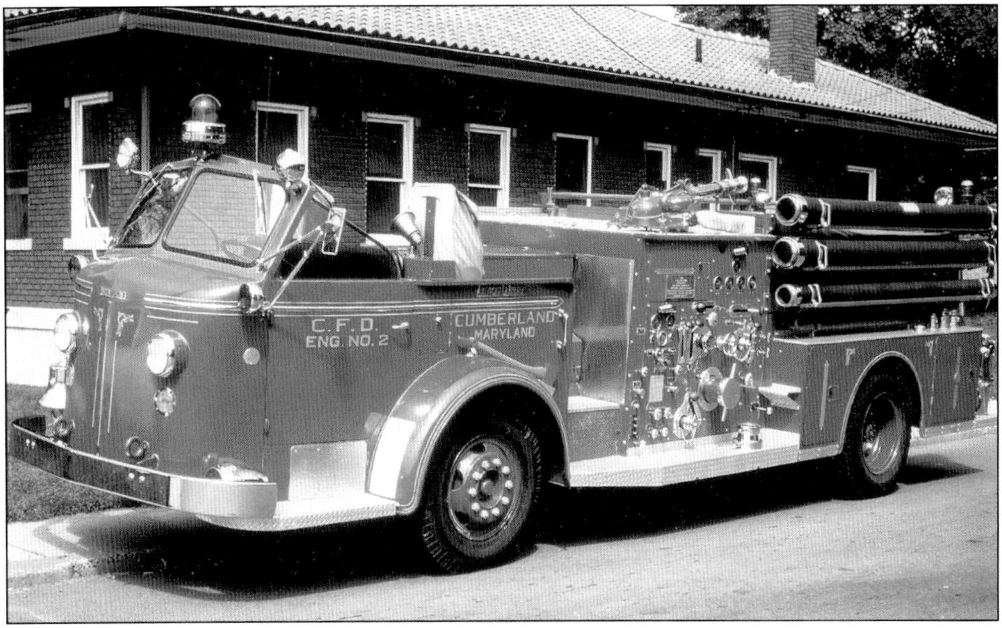

In 1958, American LaFrance delivered this 800 series pumper to CFD at its Central Station. It could pump 1,250 GPM and carry 300 gallons of water. It was eventually placed in Constitution Park in place of the 1931 Ahrens-Fox.

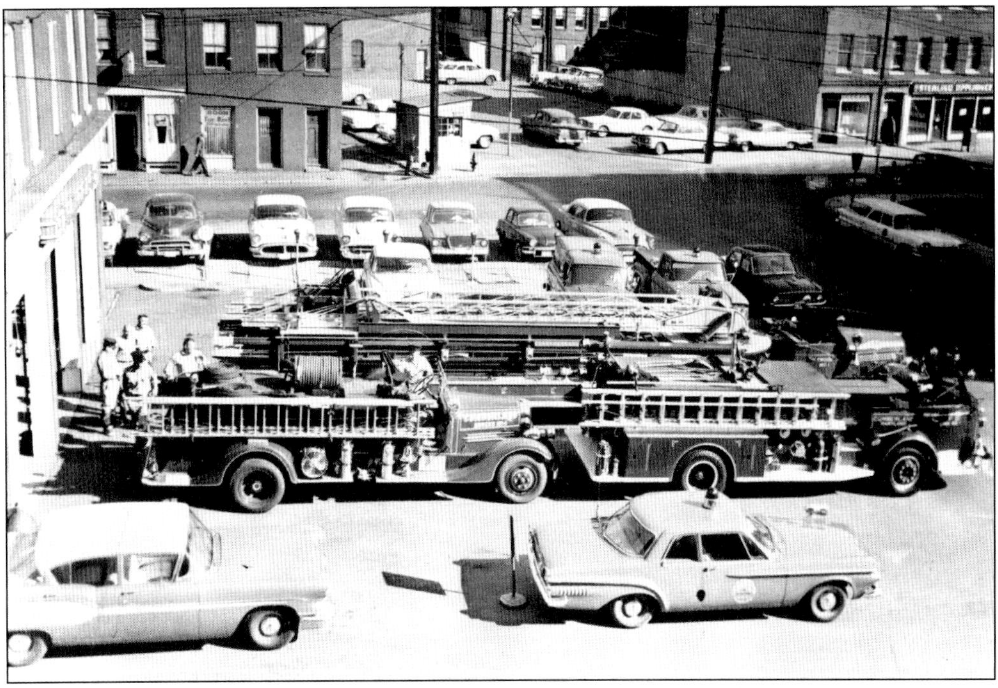

This aerial view of the Central Station apparatus fleet was taken during the early 1960s. Interesting sights include the pre–urban renewal buildings along Bedford and Centre Streets in the background. Visible apparatus include the 1958 American LaFrance pumper (far right) with an Ahrens-Fox pumper to the left. Behind the pumpers is the 1949 Seagrave aerial ladder, and in the foreground is a Dodge Dart chief's car. (Courtesy Cumberland Fire Department.)

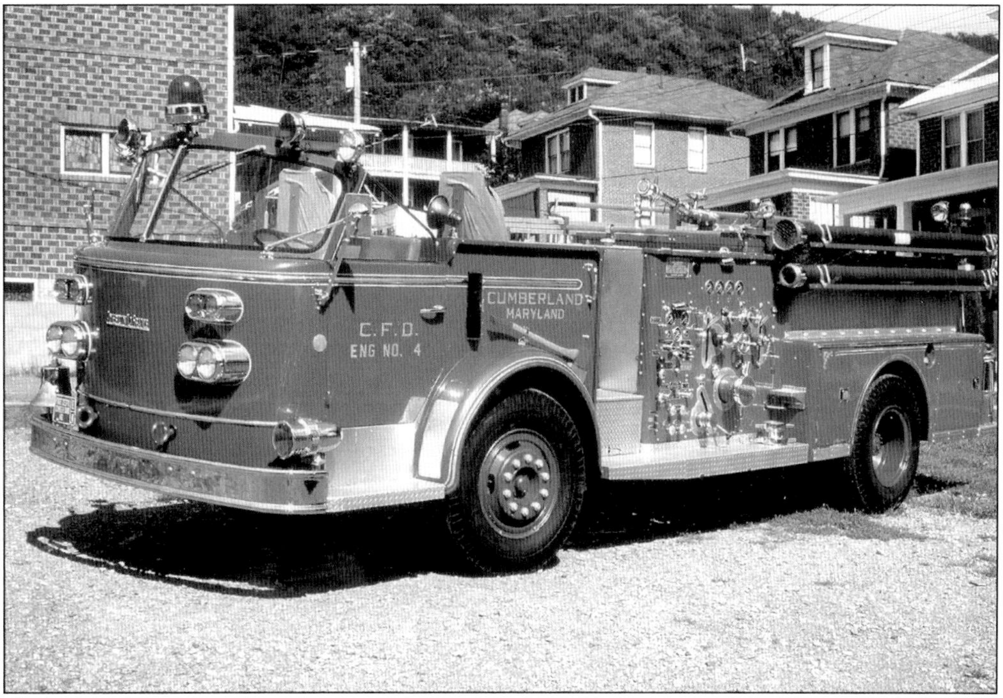

American LaFrance delivered a 1962 900 series pumper with a 1,000-GPM pump and a 300-gallon water tank. It would eventually be sold to the Mettiki Coal Mine complex in Garrett County, Maryland.

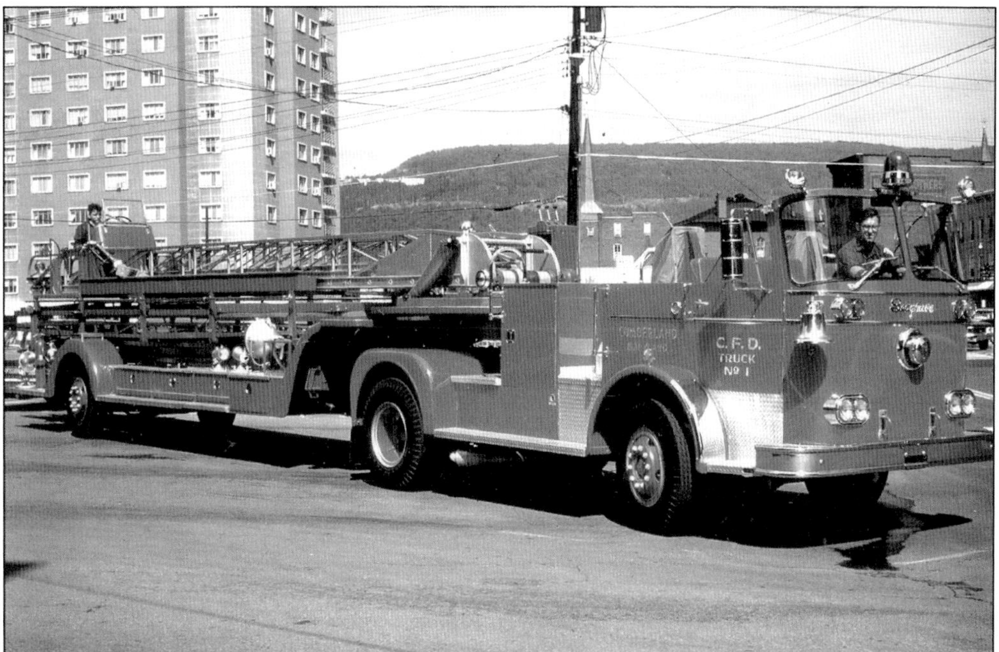

In 1968, Seagrave delivered an open-cab K model tractor to pull the 1949 Seagrave 85-foot aerial ladder trailer. Jim McDonald is the fireman driving the new tractor.

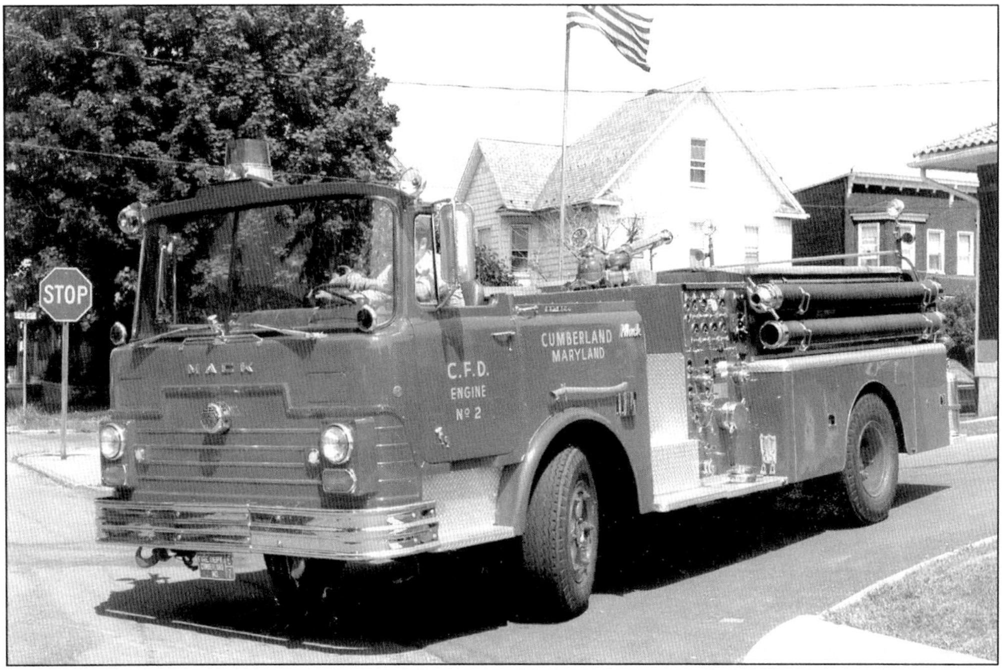

A second delivery in 1968 was this Mack CF model diesel pumper that could pump 1,250 GPM and carried 500 gallons of water. (Courtesy Virginia McLaughlin.)

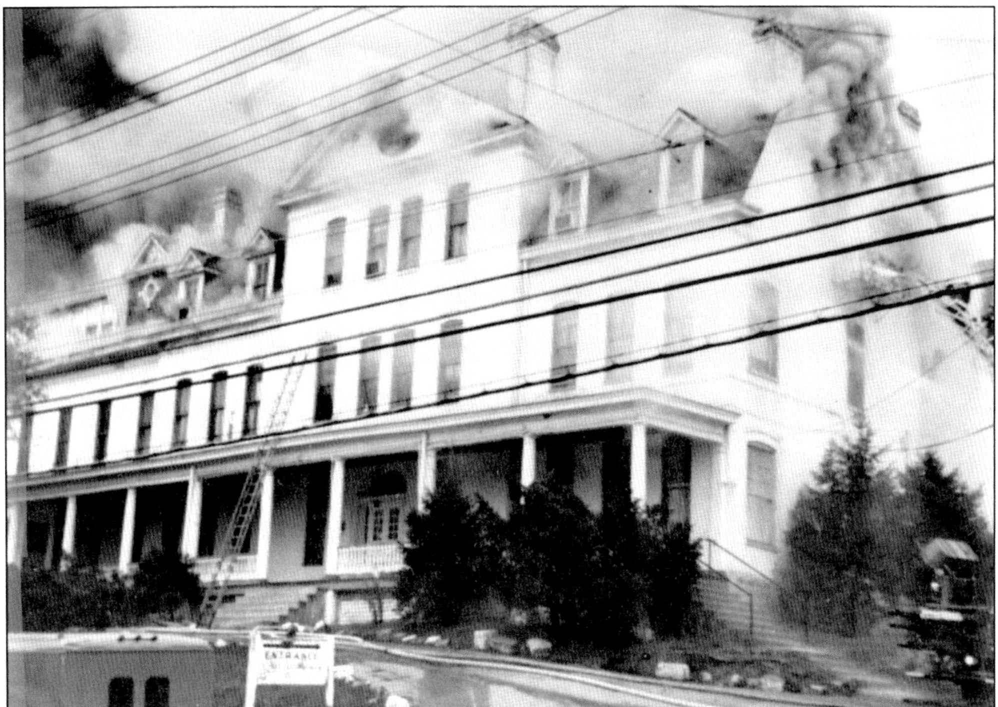

Many heroic rescues were made at the Allegany Inn fire on April 23, 1972. The building, which was originally a hospital, was serving as low-cost housing when fire destroyed the structure and killed three people. (Courtesy Cumberland Fire Department.)

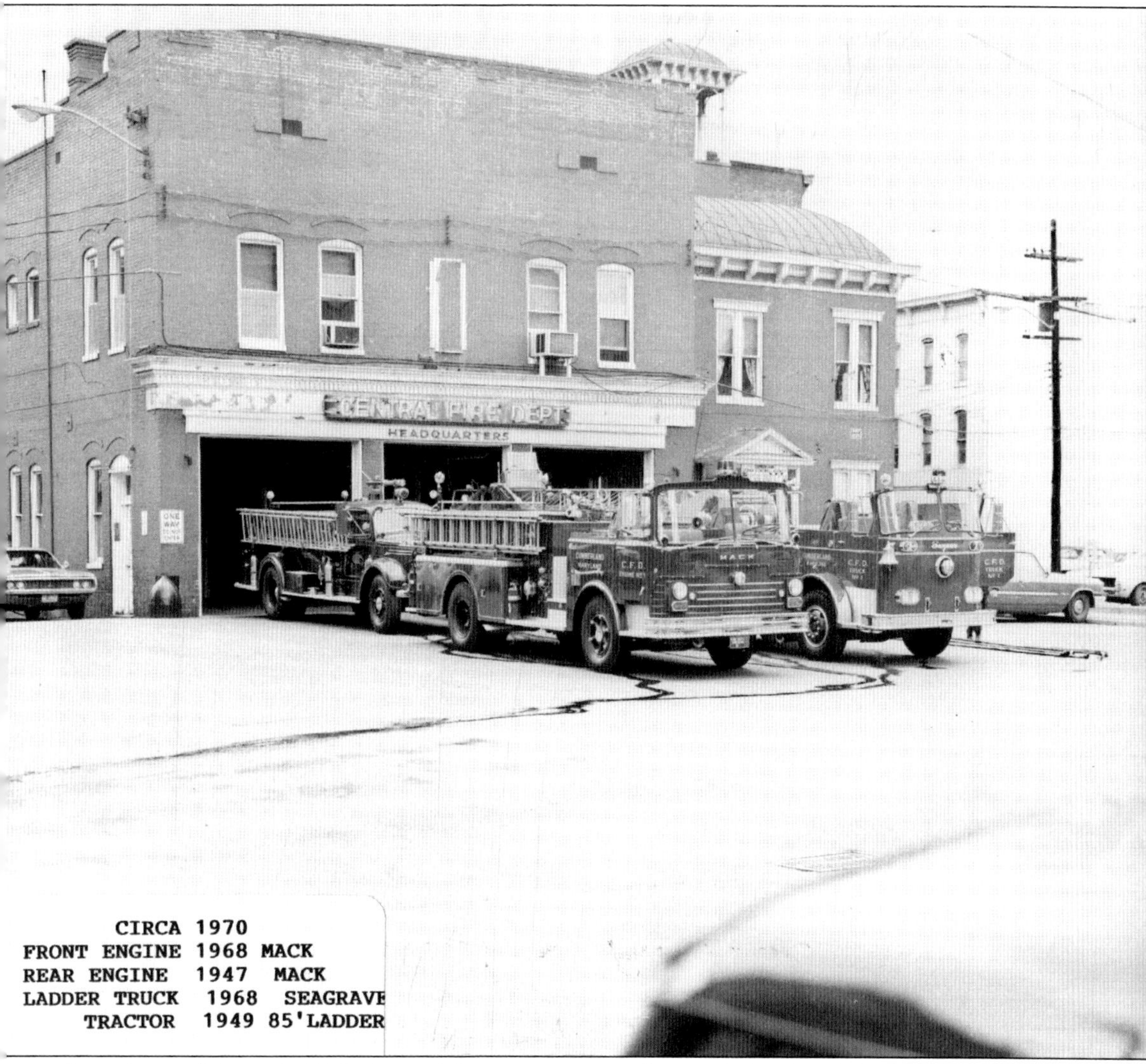

This photo of the early 1970s Central Station apparatus fleet includes the 1968 Mack CF pumper (center) with a 1947 Mack E model pumper behind it. The 1947 Mack was purchased used from another fire department to serve as a reserve pumper. The 1968/1949 Seagrave aerial ladder is also visible. (Courtesy Cumberland Fire Department.)

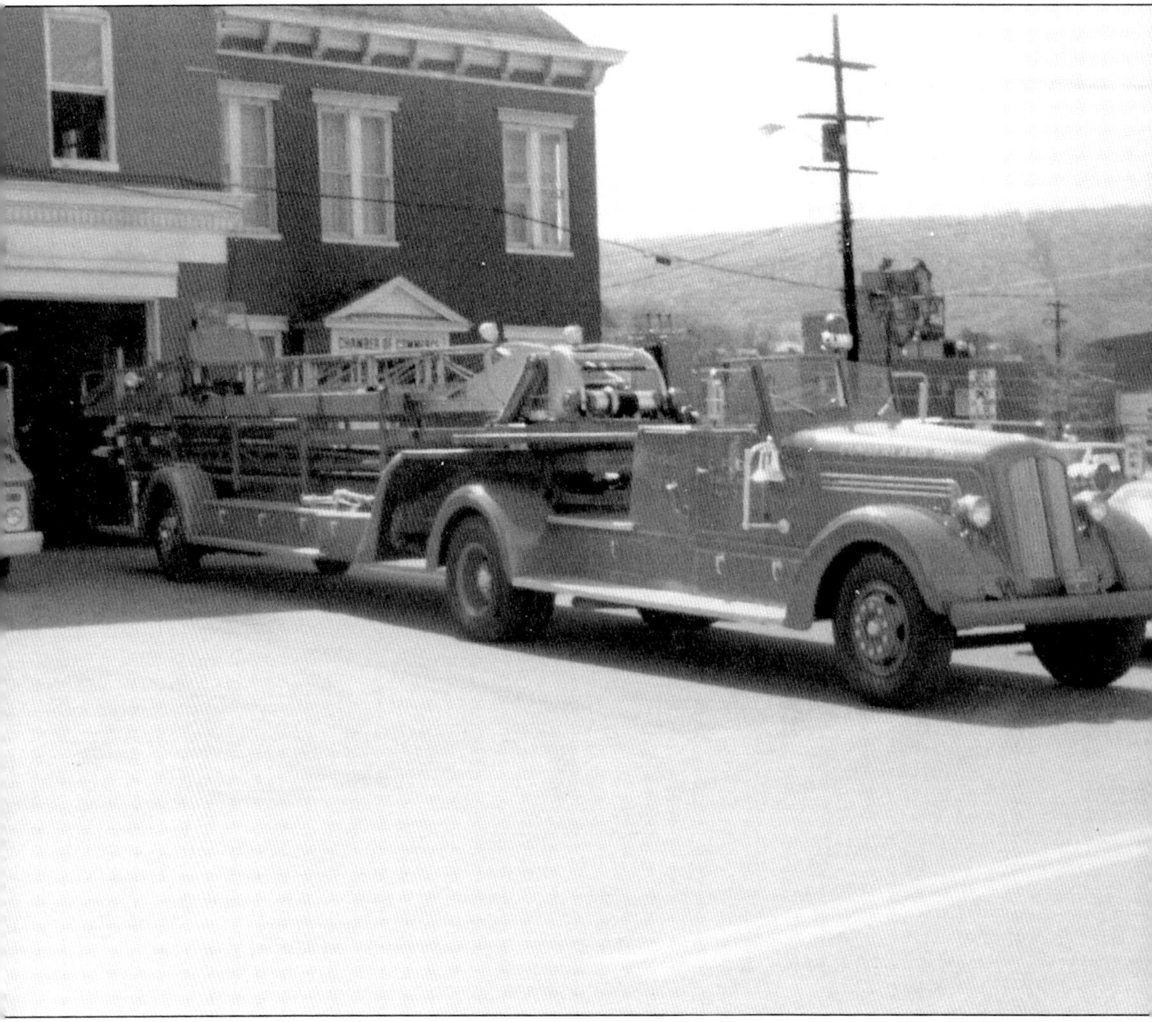

In 1972, CFD purchased this used 1945 Seagrave aerial ladder from Easton, Maryland. The 100-foot aerial was sold to LaVale Volunteer Fire Department in 1976. (Courtesy Russell Livengood.)

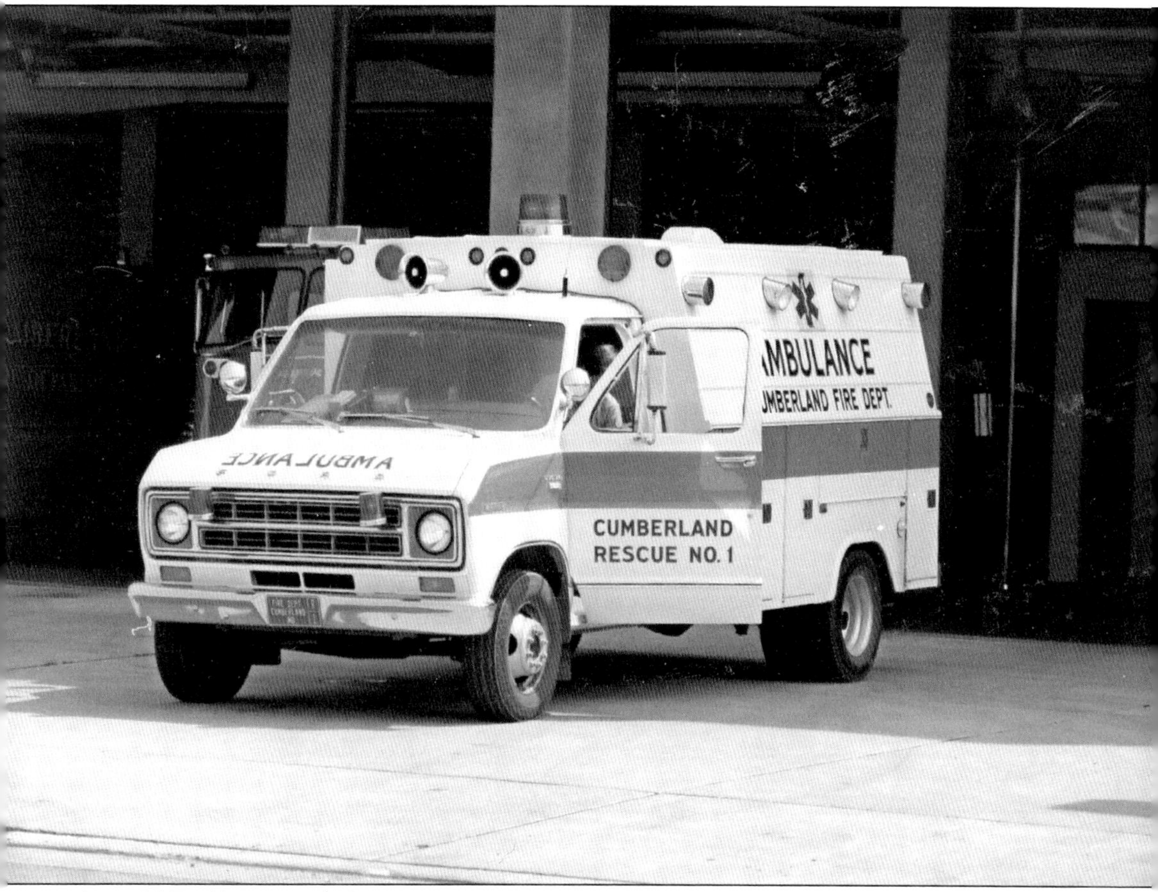

This 1975 Ford/Braun ambulance went into service in May 1977 as CFD's first ambulance vehicle. (Courtesy Russell Livengood.)

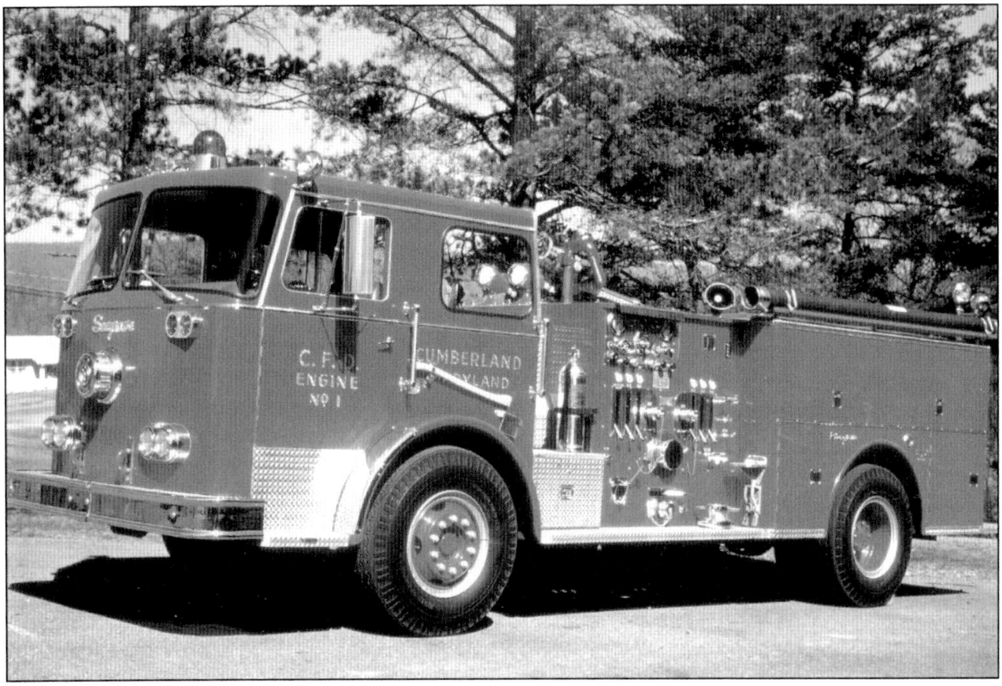

In 1976, Cumberland received this Seagrave PB series pumper, which had a 1,250-GPM pump and carried 500 gallons of water. It would be rebuilt in 1985 by Allegheny Fire Apparatus.

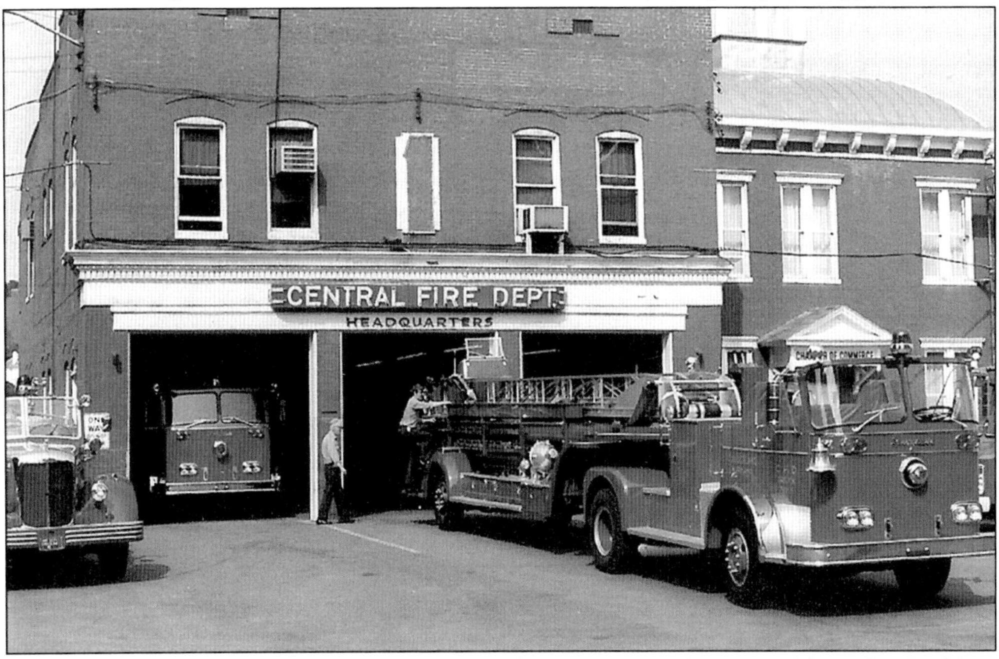

By 1977, Central Fire Station was reaching the end of its service. The new Public Safety Building was nearing completion, and CFD members were looking forward to new quarters. The apparatus visible in this photograph from left to right are a 1950 Mack L pumper, 1976 Seagrave pumper, 1968/49 Seagrave aerial, and the 1968 Chevrolet step-van. The members walking to the ladder truck are Paul Borror (left) and Gary Shanholtz. (Courtesy John Floyd.)

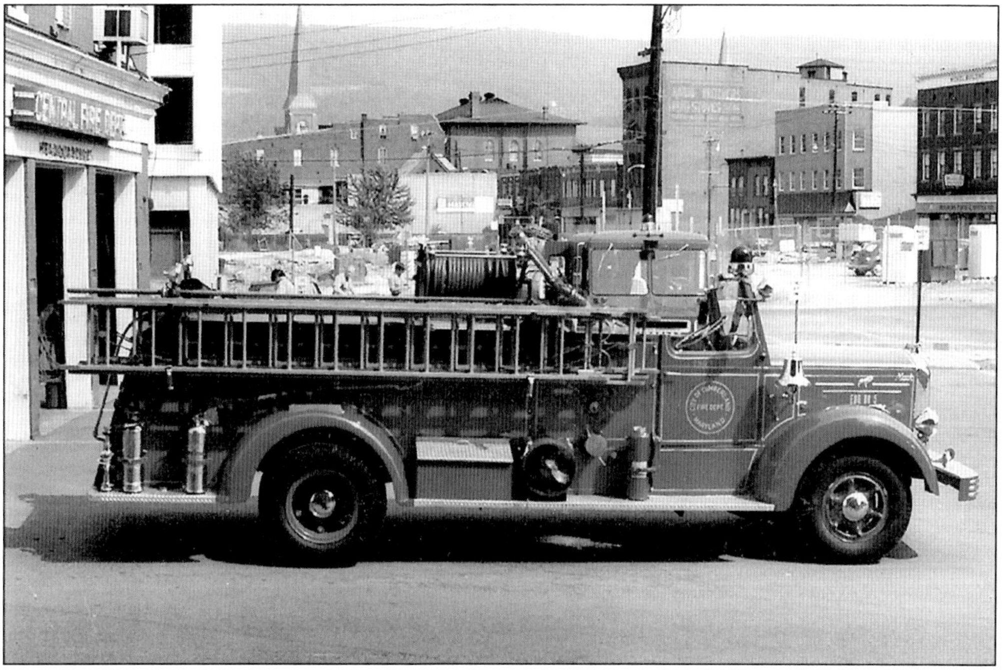

In this 1977 view, the 1950 Mack still looks good after 27 years of service. It is still on the CFD roster today. Also note the new Public Safety Building behind the apparatus and the large amount of demolition of neighboring buildings. (Courtesy John Floyd.)

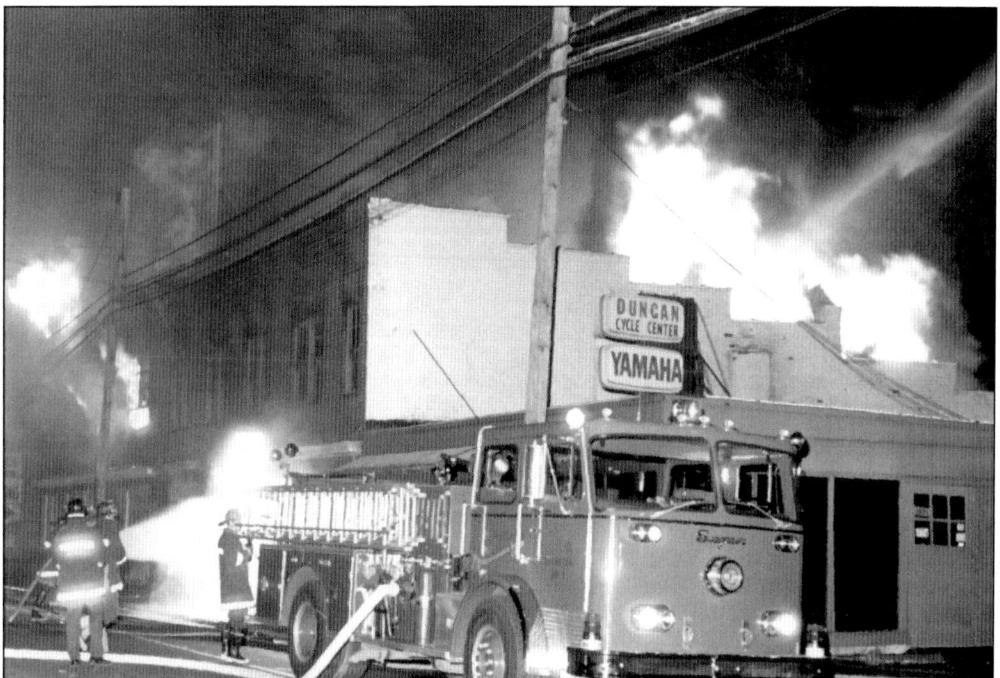

In 1979, the Duncan Cycle Shop on North Mechanic Street burned in a hot, fast-moving fire fed by the crates housing dozens of motorcycles. The 1976 Seagrave pumper shown suffered damage from radiant heat. (Courtesy Cumberland Fire Department.)

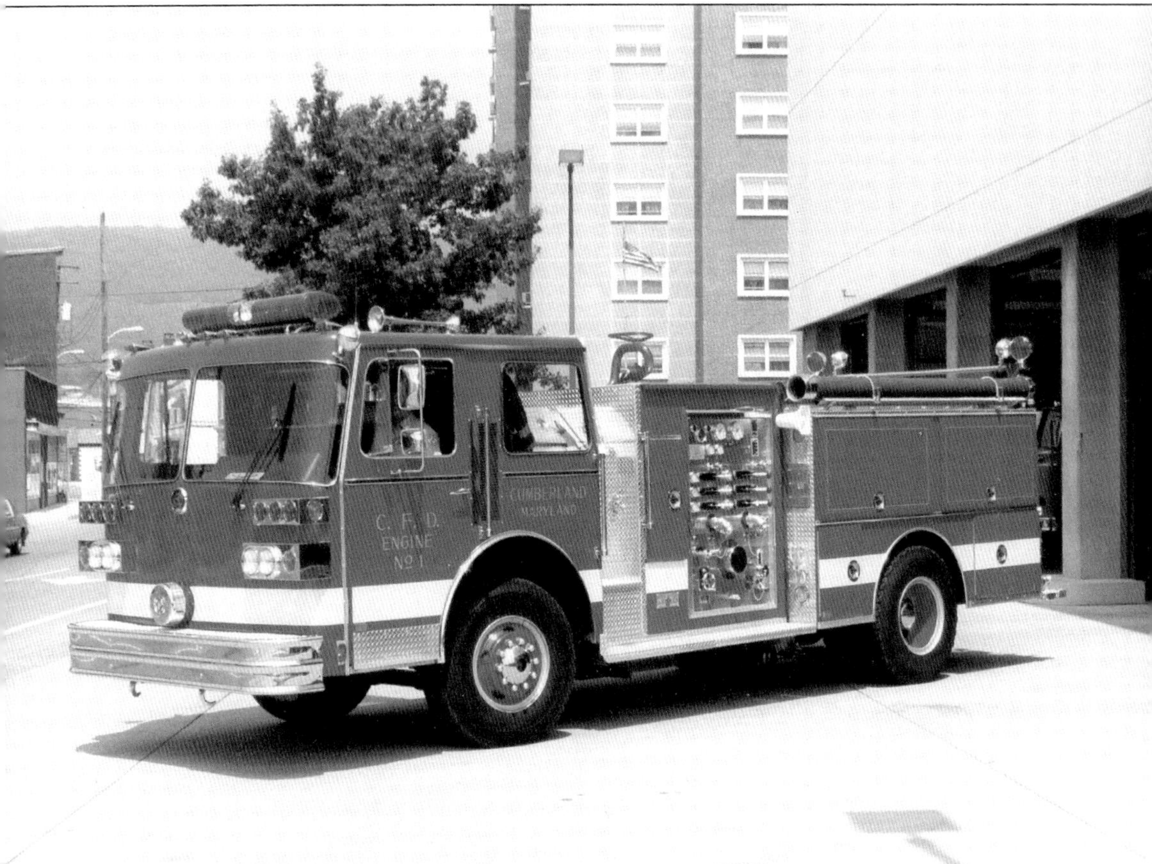

CFD received a new 1987 Sutphen pumper that could pump 1,250 GPM and carry 500 gallons of water. Sutphen, a manufacturer based in central Ohio, has made significant deliveries to Cumberland and other nearby communities in recent years. (Courtesy David Bowen.)

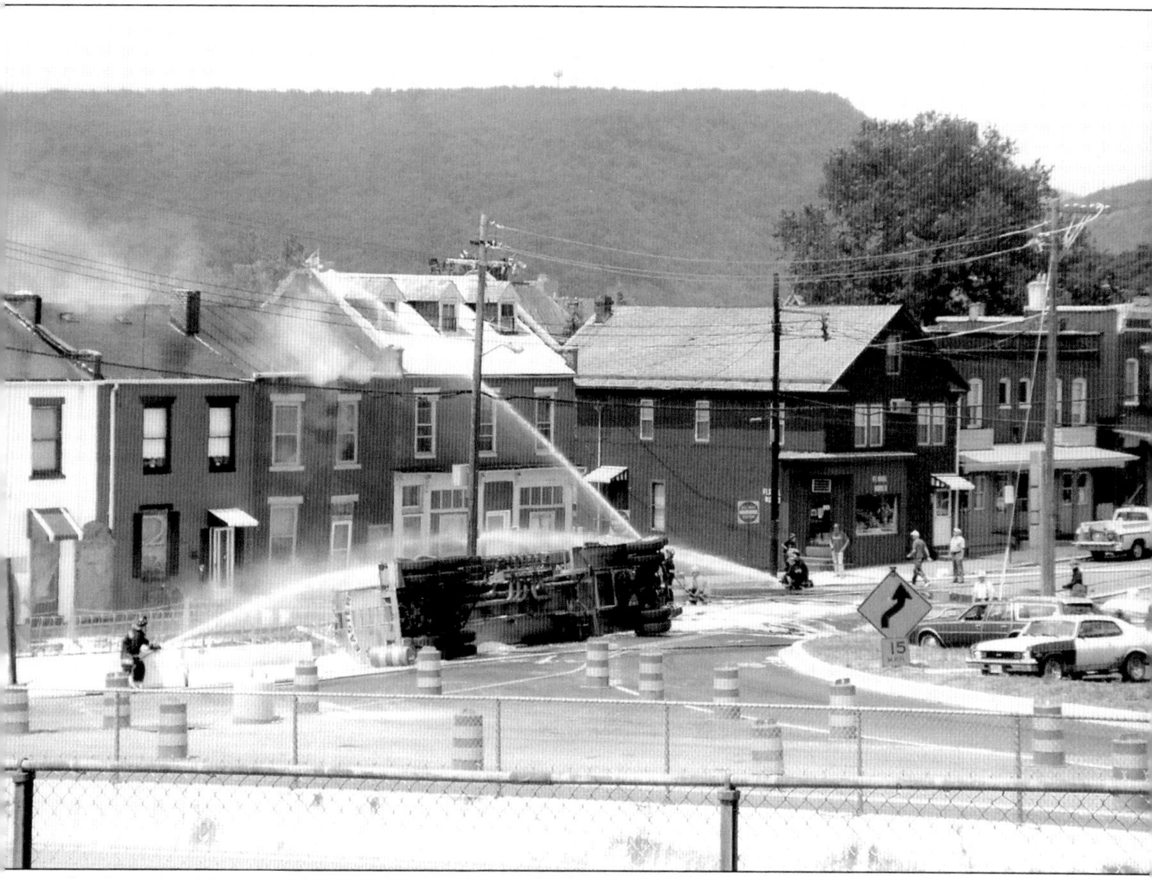

On June 1, 1991, a gasoline tanker truck wrecked on Interstate 68 at the Maryland Avenue exit. The gasoline poured into the basements of the adjacent homes and eventually ignited, destroying several buildings. This view shows firefighters trying to smother the tanker with foam as a house begins to show smoke. (Courtesy David Bowen.)

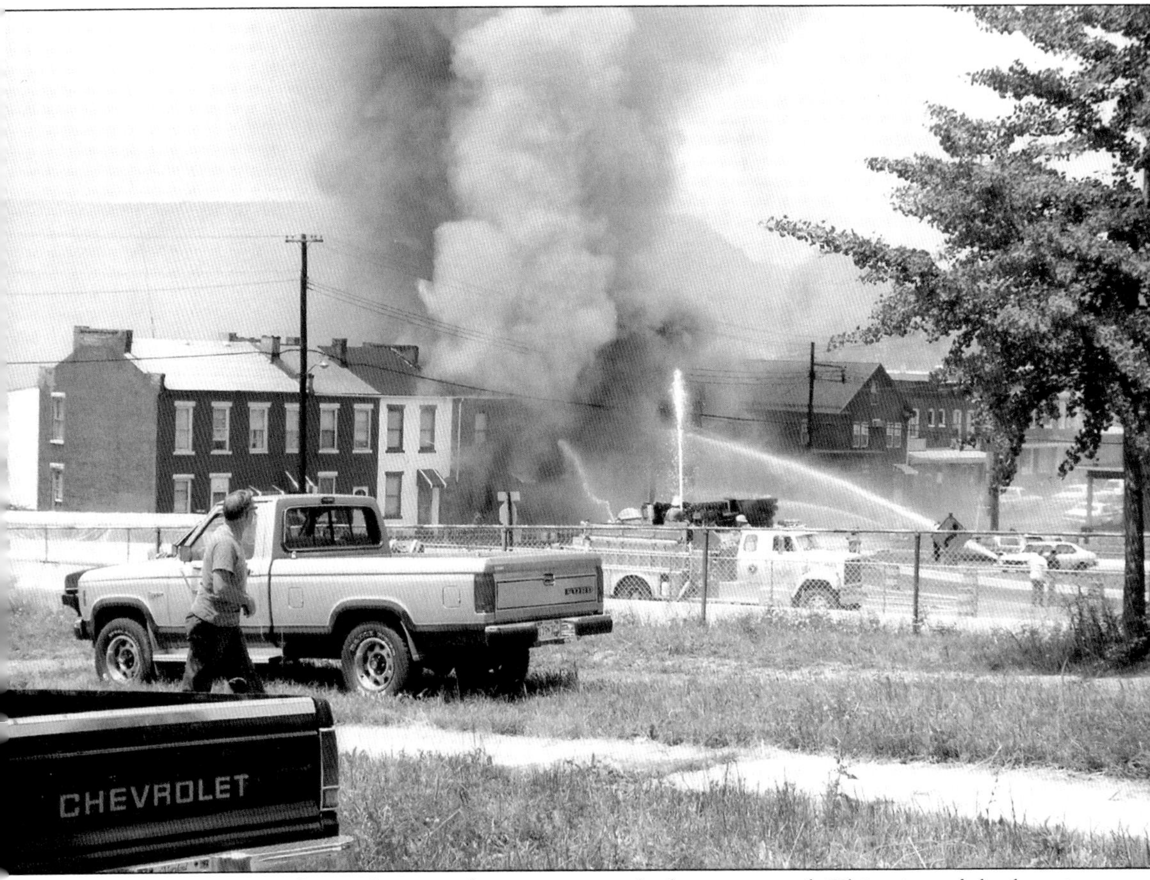

After the first explosion from gas fumes, many onlookers scattered. This view of the burning houses shows a pumper from District 16 VFC, which was one of the many volunteer mutual-aid units that responded. (Courtesy David Bowen.)

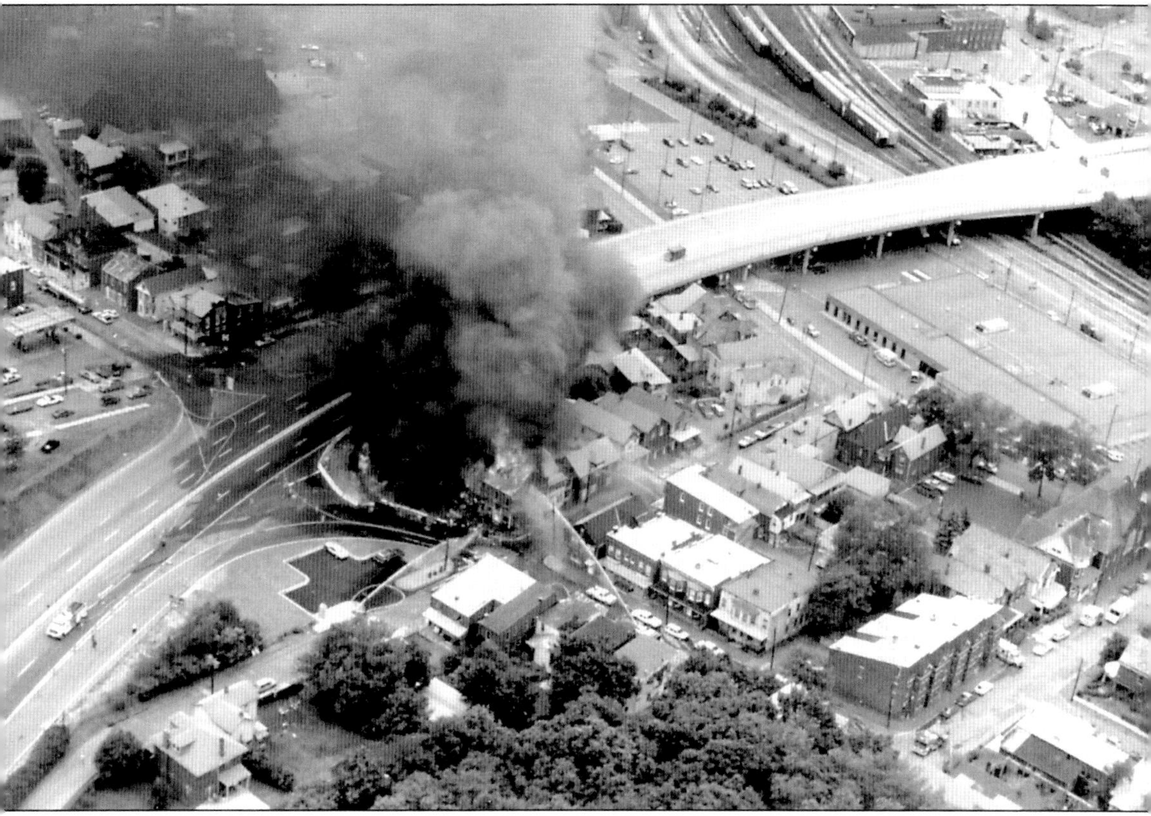

Here is an aerial view of the June 1, 1991, gasoline tanker fire. When the spilled gasoline ignited the entire block of houses, the airplane was probably the safest place to be.

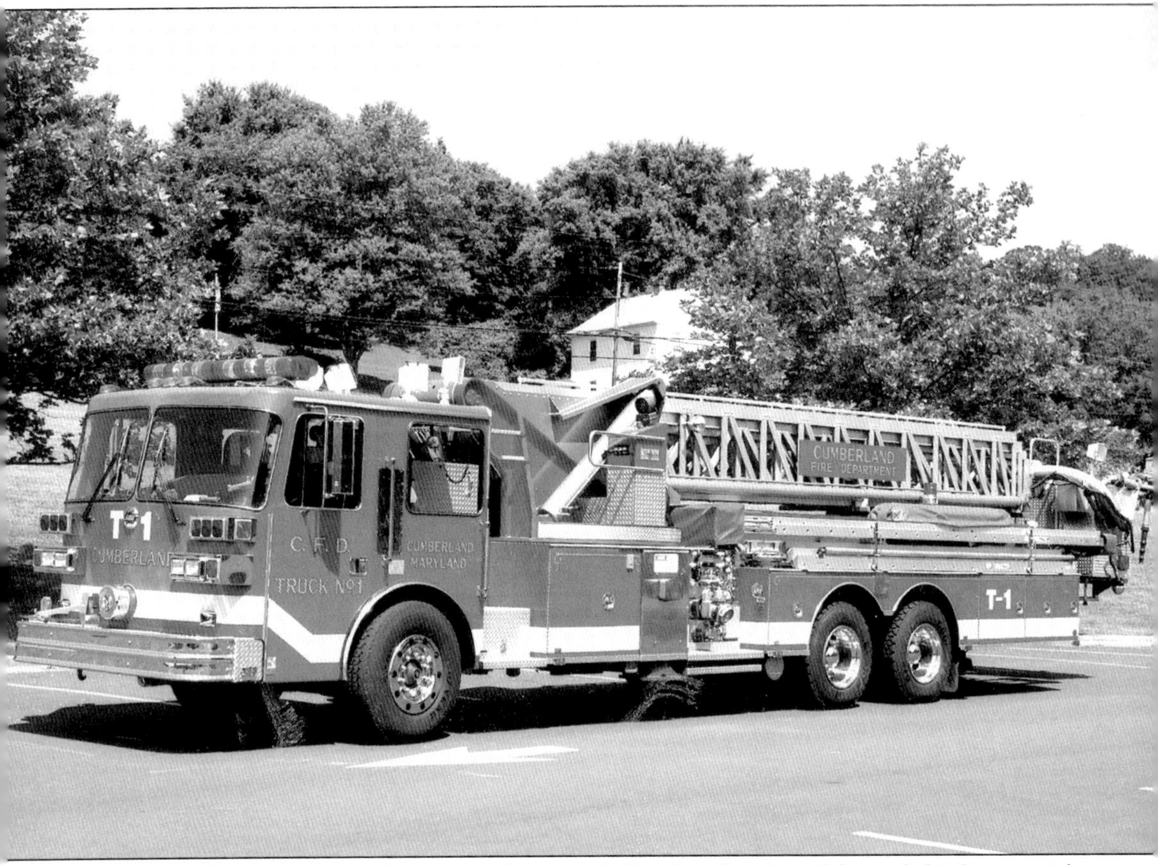

CFD finally obtained a modern aerial device in 1992 when Sutphen delivered this big rig with a 1,750-GPM pump, a 300-gallon water tank, and a 90-foot aerial tower.

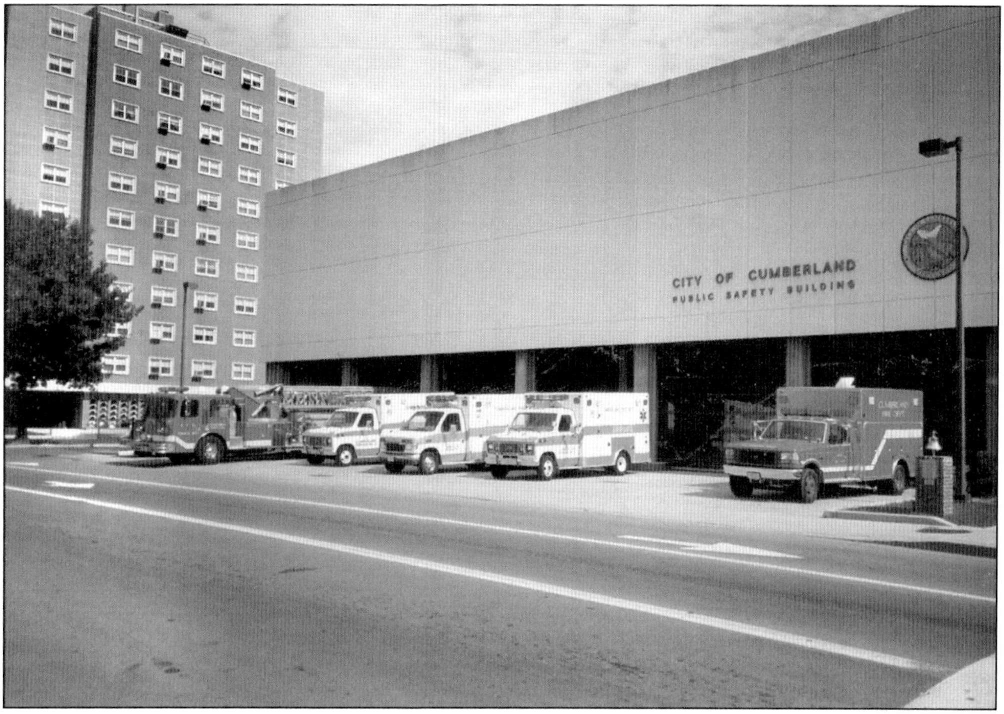

This 1993 view of the Public Safety Building Headquarters shows, from left to right, new Tower One (a 1992 Sutphen), three ambulances, and new Rescue Squad 1 (a 1993 Ford/4 Guys).

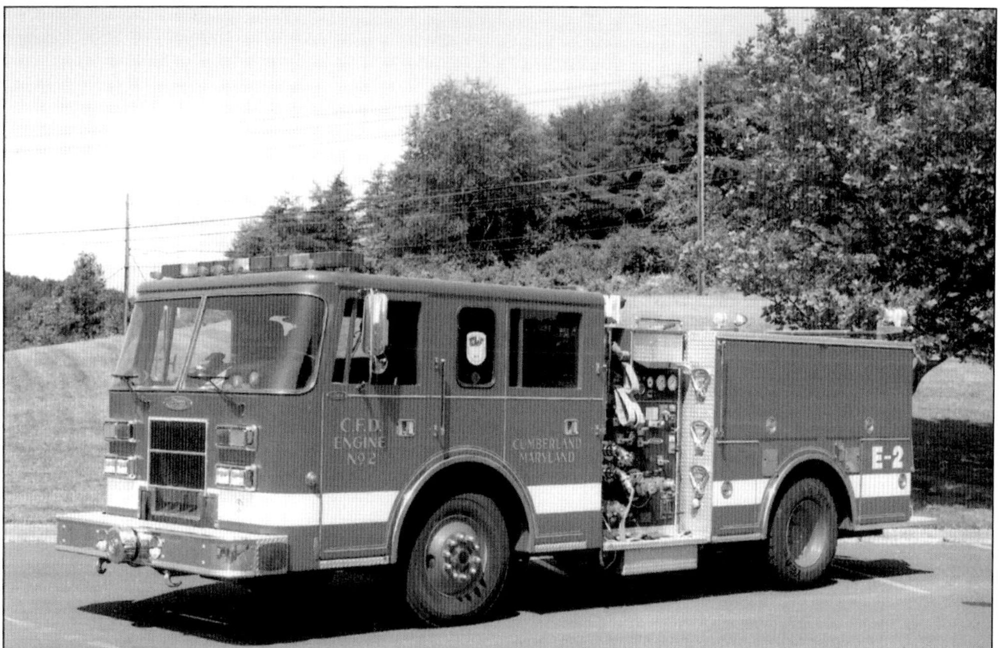

Pierce delivered a Saber model pumper with a 1,250-GPM pump and a 500-gallon water tank in 1944. This apparatus is identical to units built for Baltimore, Maryland, at the same time, and CFD used this fact to good advantage for the city's finances.

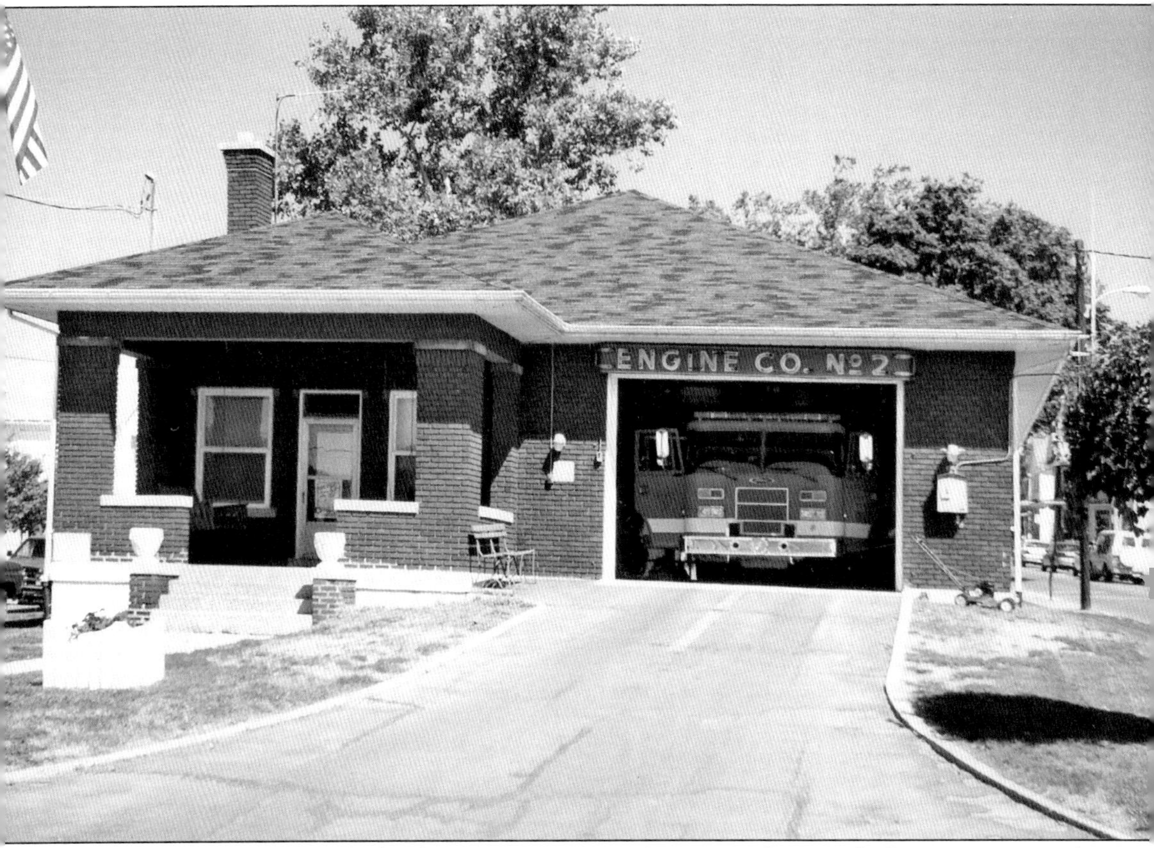

Station 2 in South Cumberland is nearing 80 years of service and is now the last surviving single engine company house in CFD. It retains many of its timeless qualities, including the 1950s vintage neon Engine Company No. 2 sign.

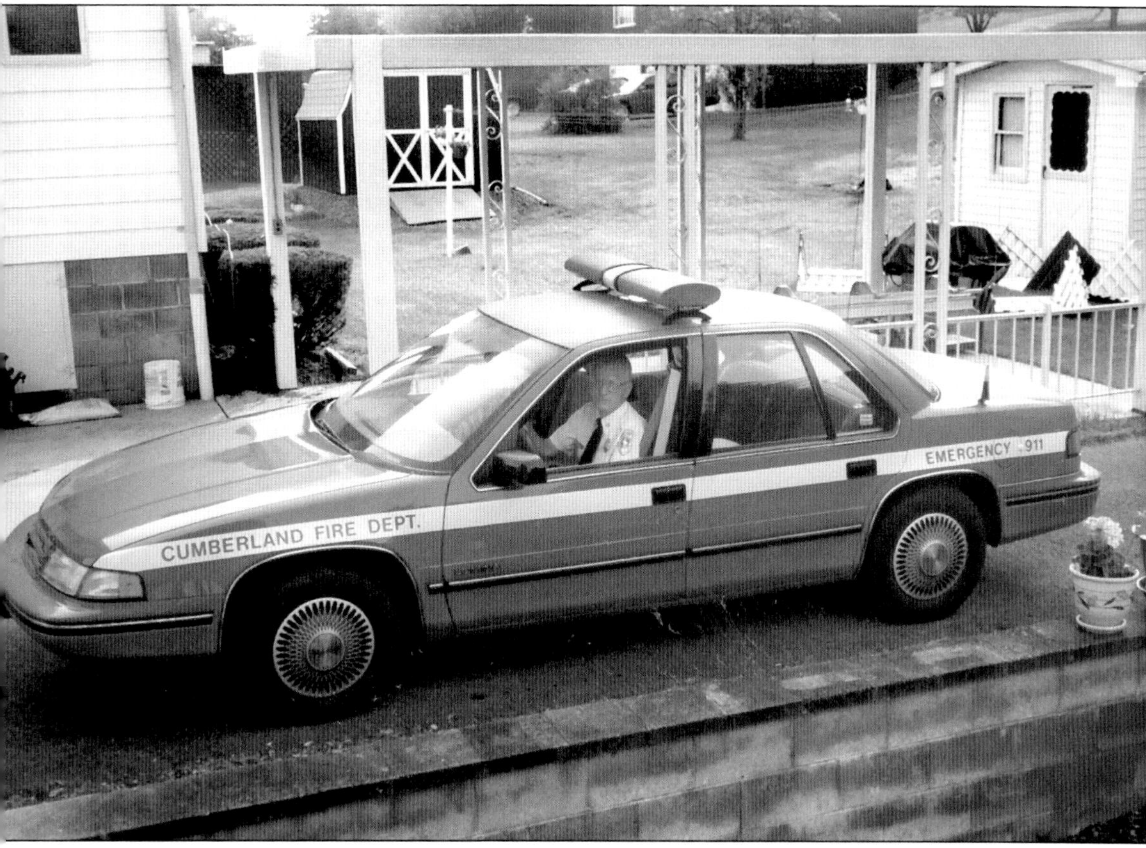

In July 1998, Chief Russell Livengood prepares to go to his last day on the job after 23 eventful years as chief and 42 years as a member of CFD. (Courtesy Russell Livengood.)

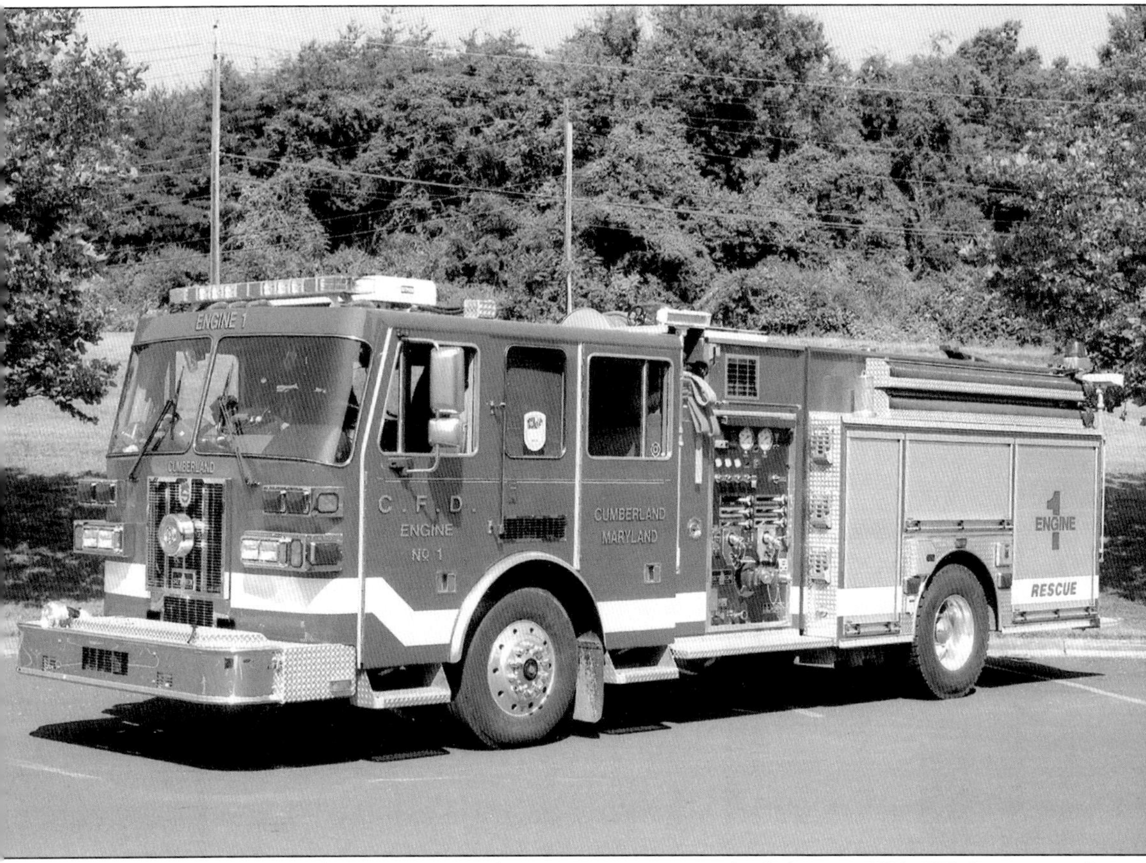

Cumberland received this 2003 Sutphen rescue pumper, which has a 1,500-GPM pump and carries 500 gallons of water. This unit can perform either as a fully equipped pumper company or as a heavy-duty rescue squad.

Two

MOUNTAIN CITY FIREFIGHTING

After two destructive business district fires in Frostburg in 1874, the local press advocated fire safety. In March 1878, the press reported on the formation of Frostburg Fire Department No. 1 and the use of the St. Michael's Church bell for alarms. By 1878, Frostburg was the second largest town in Allegany County and the commercial center of the Georges Creek coal region. By 1900, Frostburg volunteers had helped form the Maryland State Firemen's Association, traveled by train to fight blazes in other towns, and handled major fires in town plus a major cyclone in 1891.

During the early 1900s, Frostburg Fire Department (FFD) was quartered in city hall. Three other hose houses were built in outlying neighborhoods. The apparatus in 1912 included four hose reels, two ladder wagons, and a horse-drawn 1907 chemical wagon. In December 1917, a major fire on Main Street forced a reassessment of fire department needs.

The fire department's first pumper, an Ahrens-Fox RK-4 model, arrived in 1922. In 1924, the present Station No. 1 was built on South Water Street beside the city hall station. Frostburg's apparatus fleet grew with an American LaFrance quad (which can perform four functions) in 1940, another Ahrens-Fox pumper in 1949, and rescue vans in 1953 and 1958. After obtaining a 1961 Seagrave pumper and a used 1941 Seagrave aerial truck in 1964, Station No. 1 was overcrowded. Station No. 2 was built on East Main Street in 1966, and in 1967, a Dodge 4x4 pumper and a used Civil Defense REO rescue truck were acquired.

Major business district blazes occurred in 1969, 1973, and 1975. In February 1981, a runaway truck hit several cars and a three-story commercial building at the corner of East Main Street and Centre Street. The Frostburg volunteers took delivery in December 1981 of a new Seagrave aerial.

Frostburg took delivery of an International rescue truck in 1986, a new Seagrave pumper in 1988, and a refurbished 1973 Seagrave in 1990. In August 1996, another destructive fire wrecked three buildings on East Main Street. A new HME/4 Guys pumper helped fight this blaze. In February 2004, another huge fire occurred on Main Street at one of the survivors of the 1917 conflagration. The Lyric Theater building was badly damaged, but a major effort by firefighters contained the fire.

In 2005, another potentially serious Main Street fire was quickly extinguished by Frostburg firefighters using their newest apparatus, a Pierce quint displaying the longtime motto of Frostburg Fire Department No. 1, "Our Deeds, Our Reward."

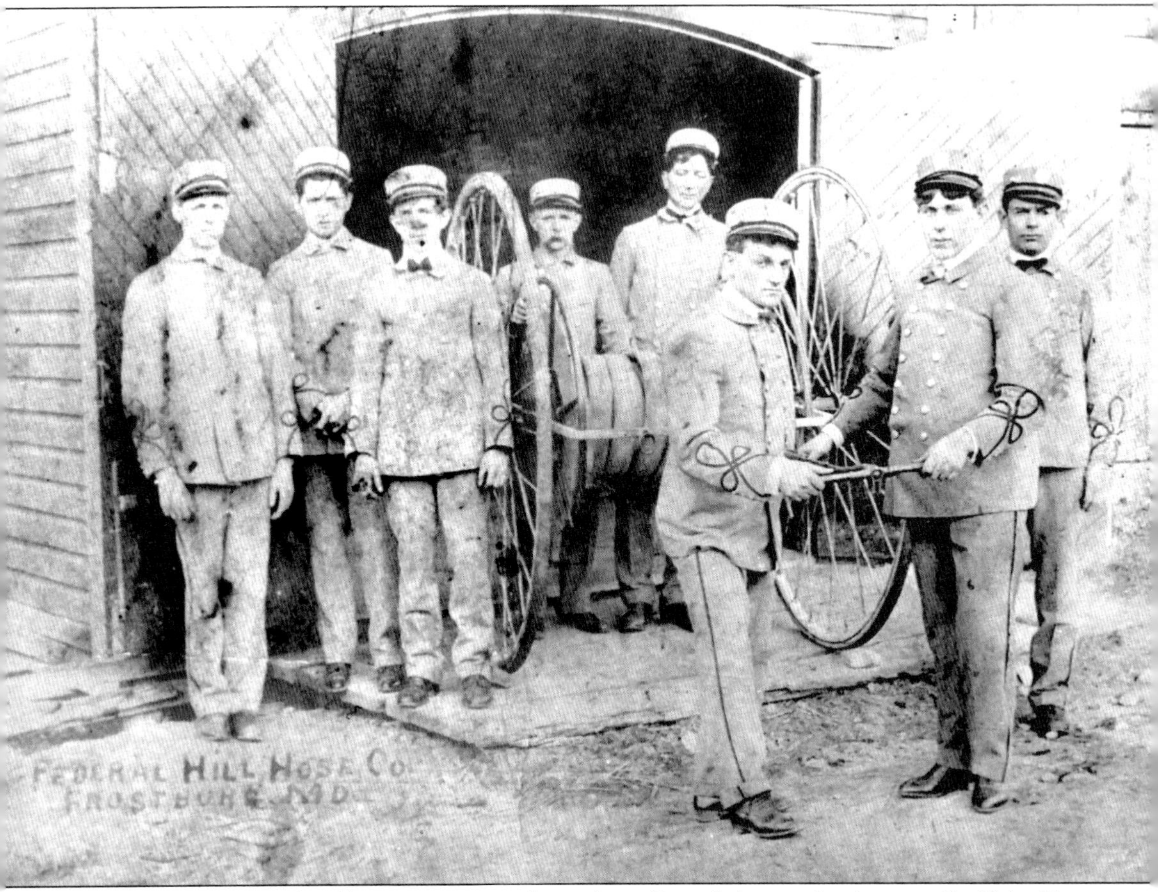

Here is the Federal Hill Hose House and members. From left to right they are Griffith, Sluss, Rase, Billy Weinke, Will Offman, Haberline, Rase, and Rank. For some years, the Frostburg Fire Department consisted of a central station in city hall on South Water Street and three outlying hose reel houses. This helped to speed the firemen's response due to the steep hills in Frostburg. (Courtesy Frostburg Museum.)

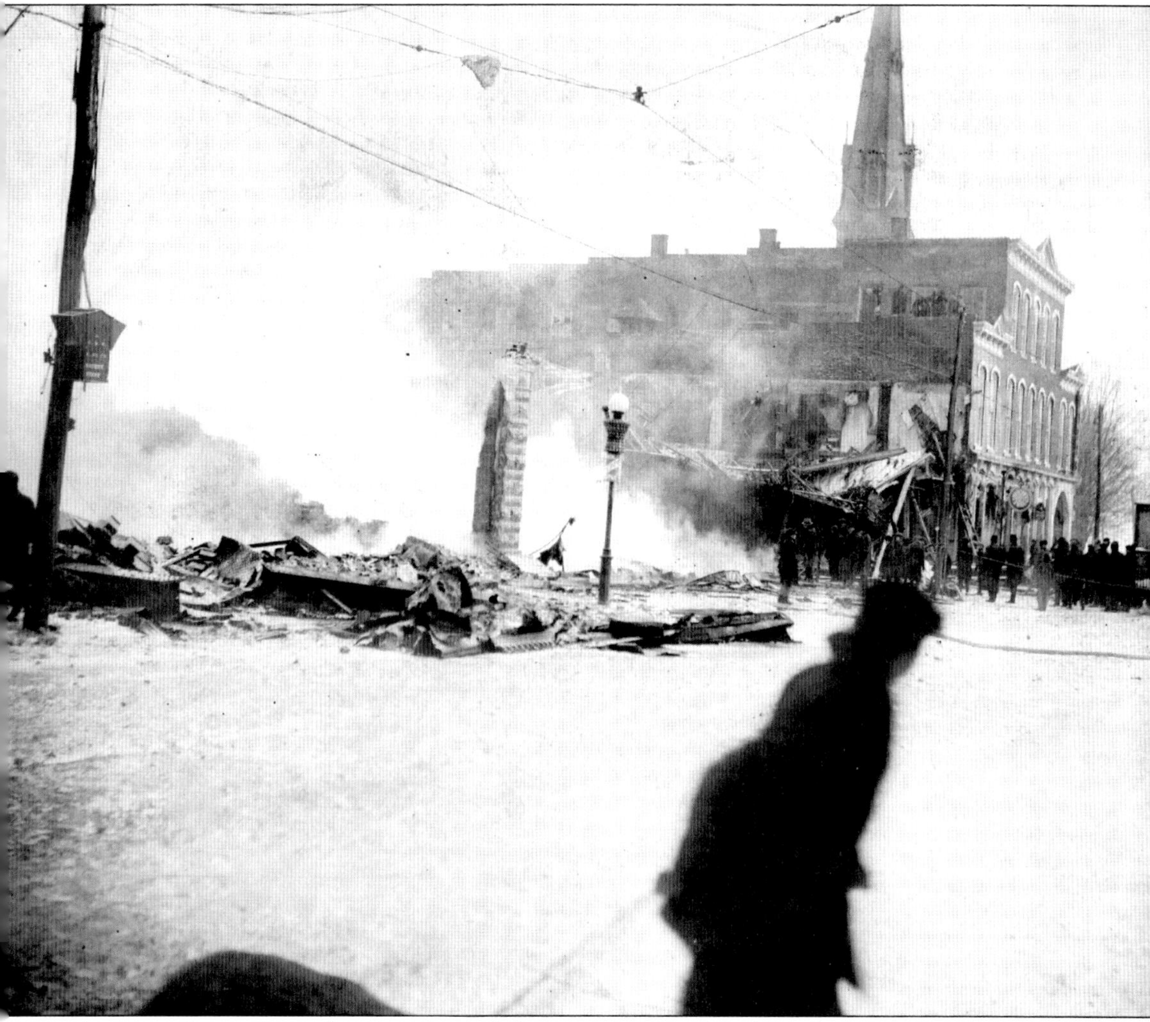

The December 1917 conflagration was a defining moment in Frostburg's history. It was a miracle that there were no deaths or that the fire did not spread farther. In this photo, the Lyric Theatre Building is seen at the right of the picture with a trolley in front. This is where Cumberland Fire Department's new American LaFrance pumper helped stop the blaze with powerful hose streams. Firemen from Midland, Lonaconing, and other nearby towns assisted Frostburg on this snowy winter day. (Courtesy Frostburg Museum.)

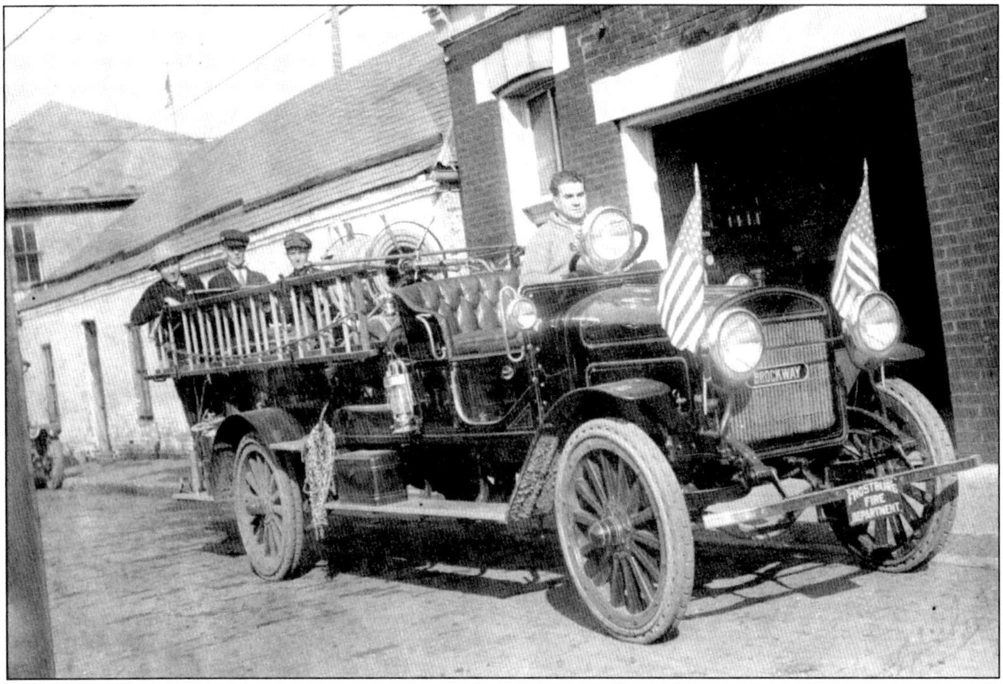

The first motorized apparatus for Frostburg was this 1917 Brockway/American LaFrance chemical and hose truck. This very old photo was probably taken around November 1918 because of the large U.S. flags on the front fenders. (Courtesy Frostburg Museum.)

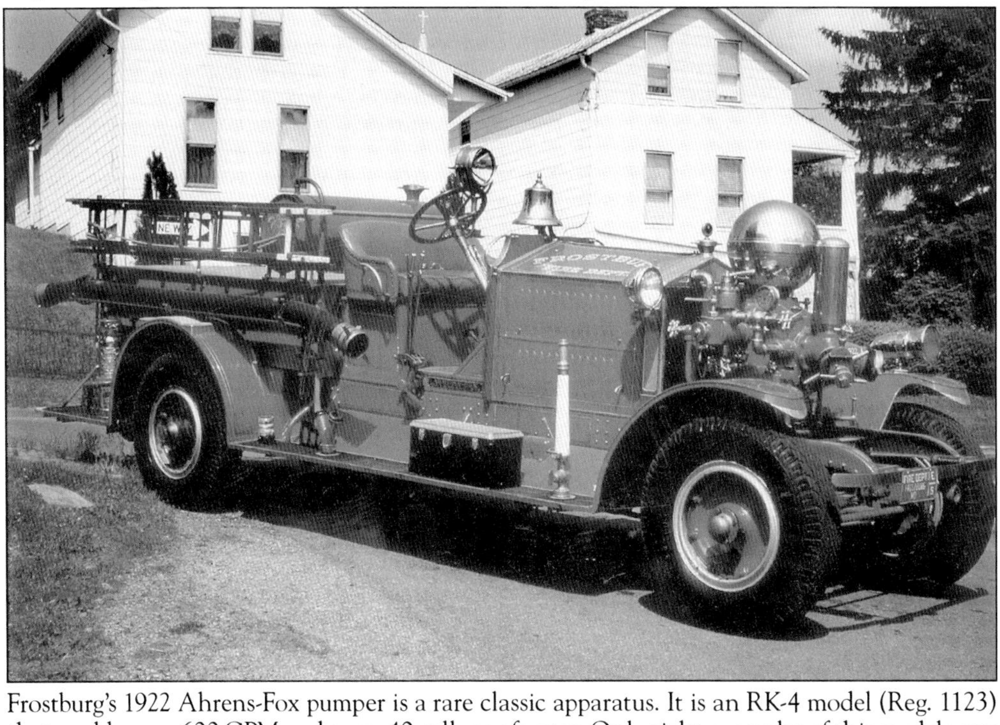

Frostburg's 1922 Ahrens-Fox pumper is a rare classic apparatus. It is an RK-4 model (Reg. 1123) that could pump 600 GPM and carry 40 gallons of water. Only eight examples of this model were built, and by 1985, five remained including Frostburg's.

The three-story Frostburg Fire Department Station No. 1 was built in 1924 and is still operating today. This photo was probably taken during a firemen's convention in the early 1960s. The city hall and police department were still located in the old fire station next door on the left.

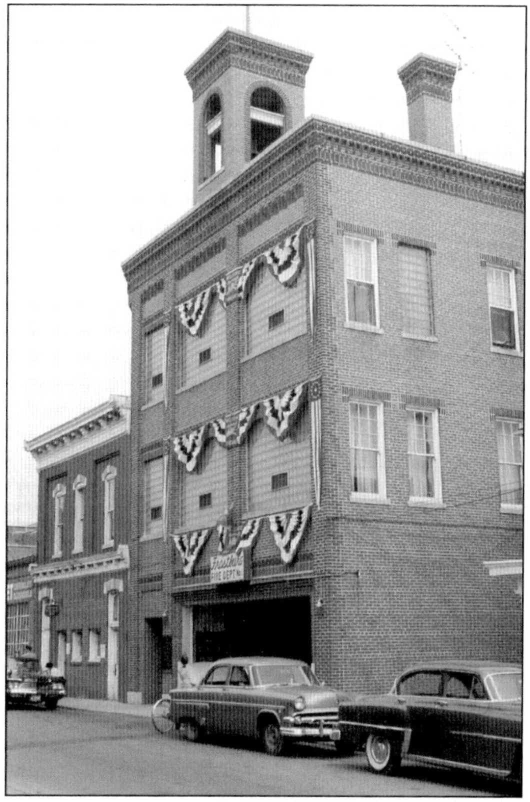

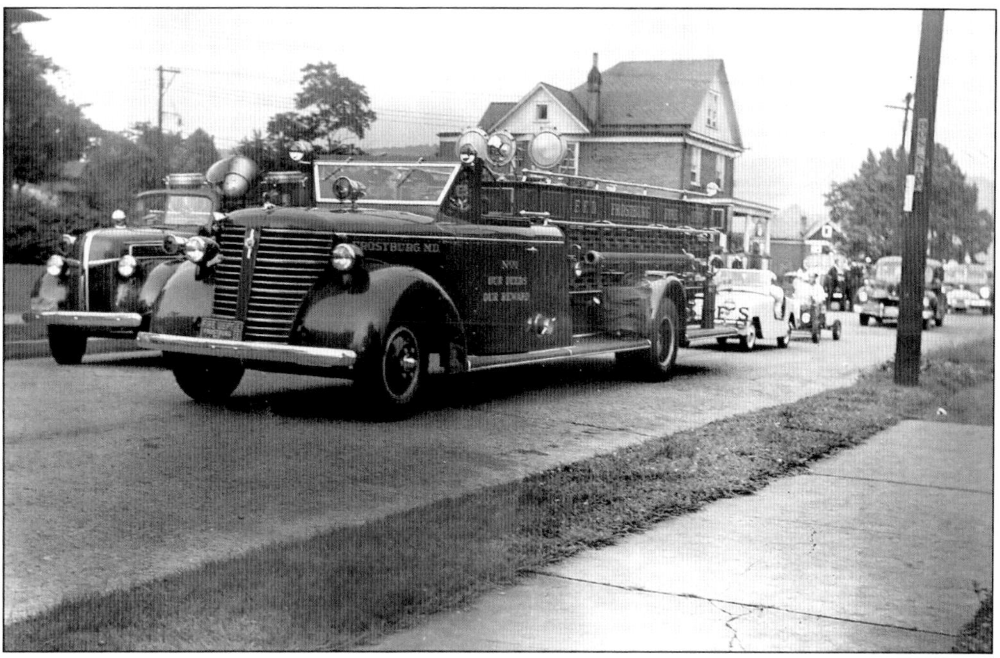

This is FFD's 1940 American LaFrance quad apparatus on parade about 1951. This truck had a 500-GPM pump, large portable ladders, hose, and a small water tank. It also had large searchlights mounted on top. (Courtesy William Snyder.)

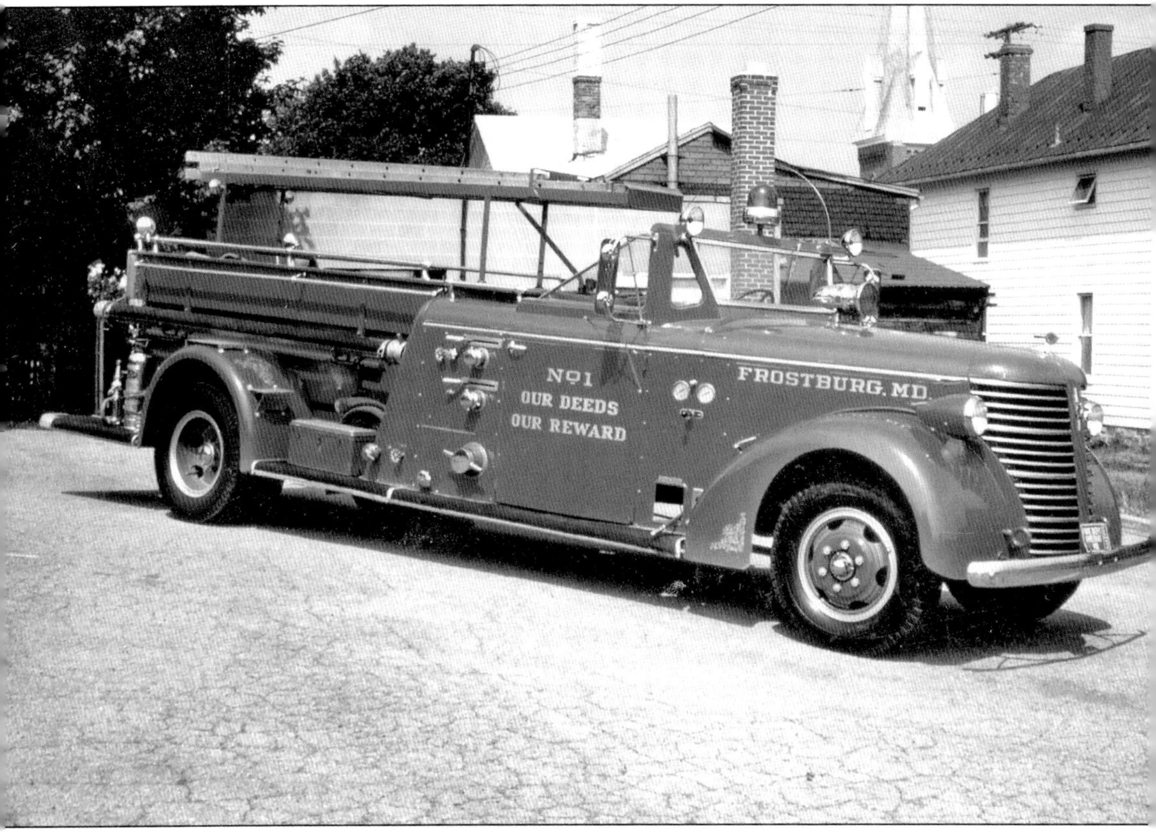

Here is the 1940 American LaFrance after it was shortened around 1968 by a local welding shop. It was modified into a regular-sized pumper. Note the ladder rack on top of the apparatus. This rig lasted until 1973, when it became privately owned and was repainted for the 1976 U.S. Bicentennial.

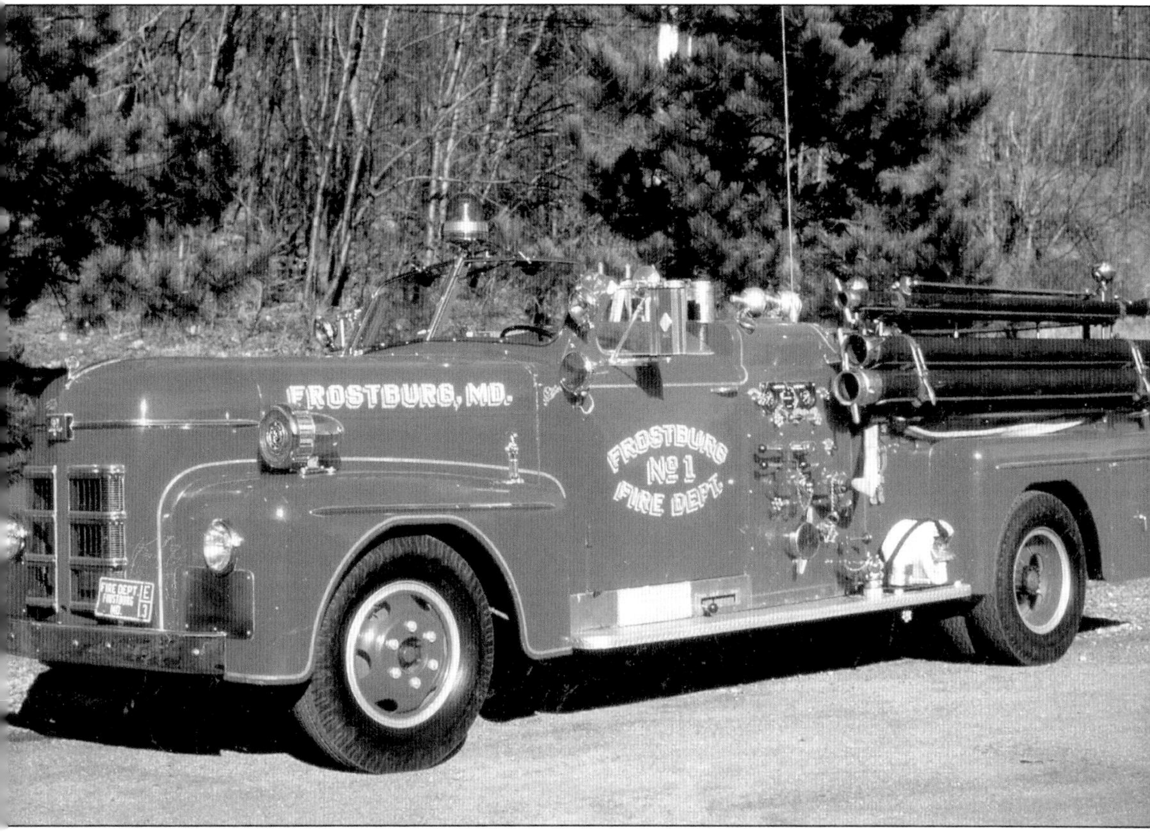

Another classic Ahrens-Fox apparatus is this 1949 EC model (Reg. 9146) that could pump 750 GPM and carry 300 gallons of water. It was sold in 1978 to a private owner in Tennessee.

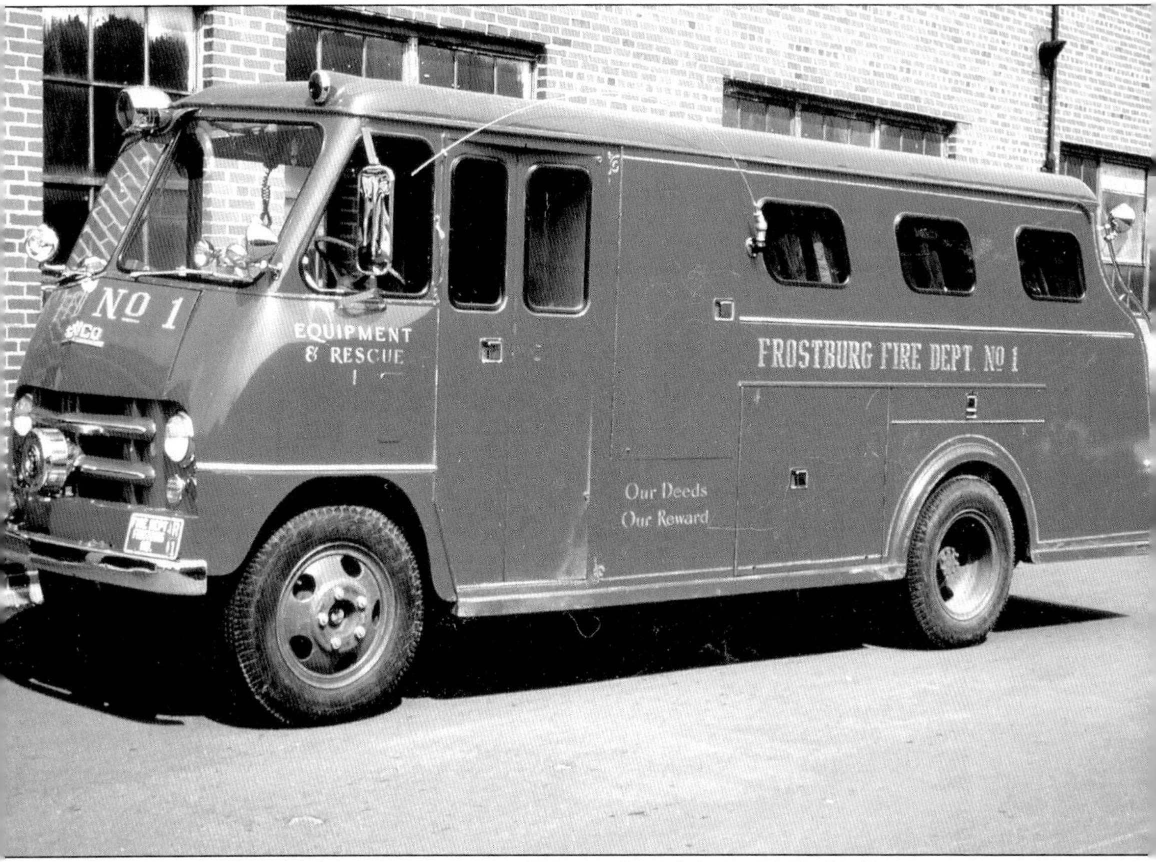

Replacing a 1953 Dodge step-van rescue, FFD purchased this 1958 Dodge/Bruco step-van rescue that also had a 350-GPM pump and carried 200 gallons of water. This versatile unit served Frostburg until 1986 and was then donated to the recently formed Rawlings Volunteer Fire Company.

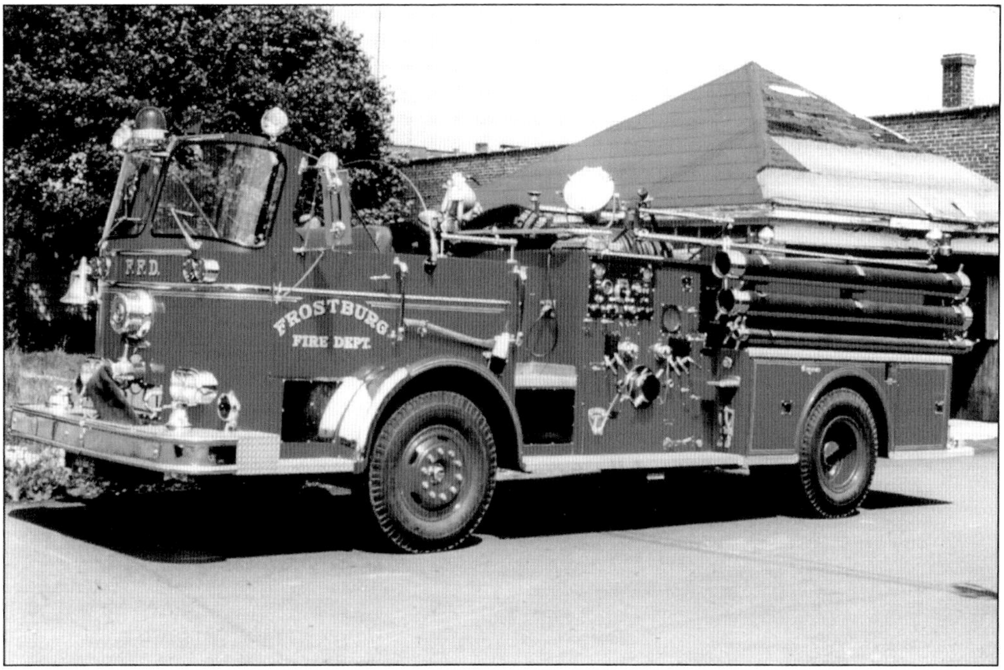

This handsome Seagrave KB model pumper was delivered to Frostburg in 1961. Pumping 1,000 GPM and carrying 300 gallons of water, this truck was updated with a Detroit diesel engine in 1981 and is now owned by veteran Frostburg firefighter Enoch Price.

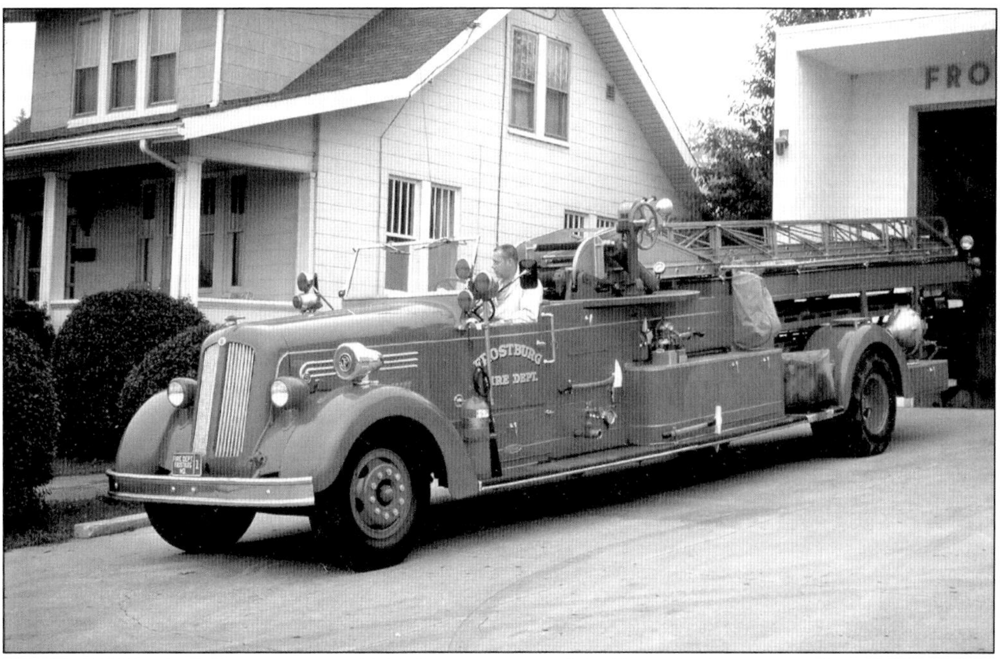

In 1964, FFD placed in service a 1941 Seagrave aerial ladder, which was purchased through a Maryland used apparatus dealer. The unit with a 65-foot steel aerial ladder was only the second such truck in Allegany County. At the wheel is Chief John C. Durst.

DEDICATION

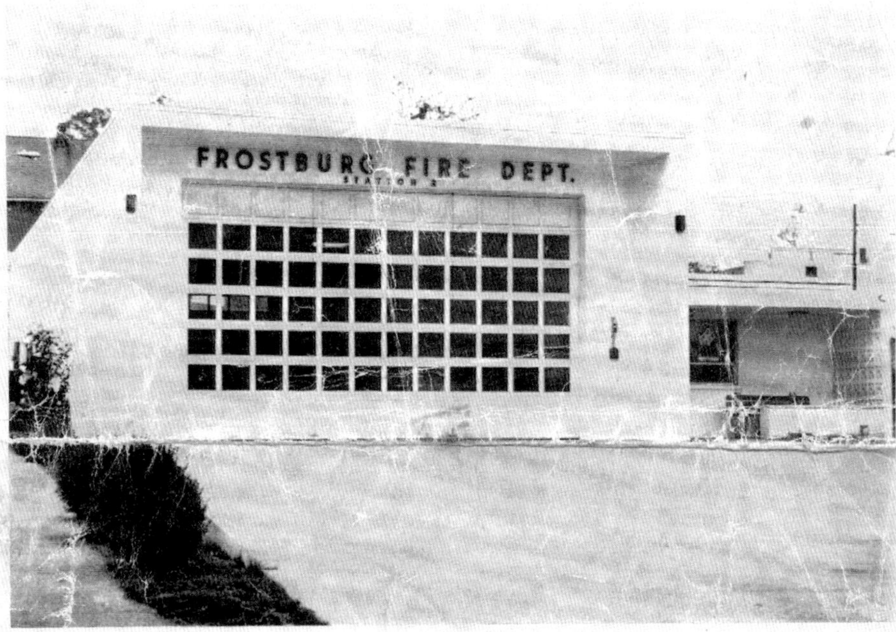

FROSTBURG FIRE DEPARTMENT

Station 2

Sunday, October 30, 1966

2:00 p.m. Rain or Shine

By 1966, a second fire station was needed in Frostburg because of the growth of both the fire department and the city. On East Main Street at Victoria Lane, this concrete "barn" firehouse was built and is an east end landmark. This is the cover of the program for the dedication, which the author attended.

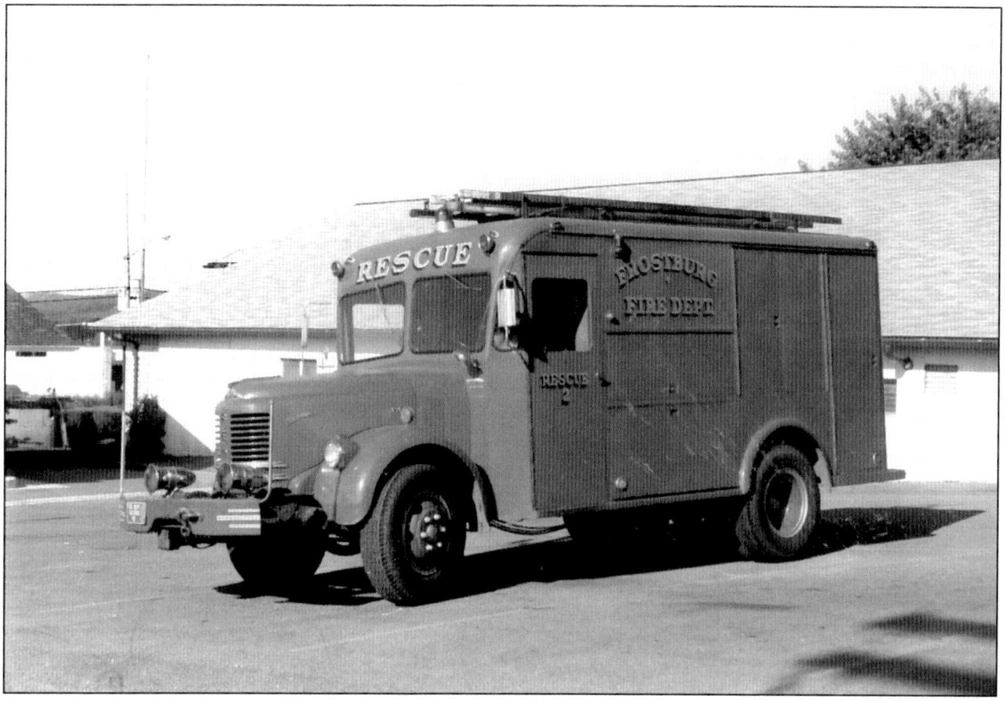

A surplus Civil Defense REO rescue truck was purchased by FFD in 1967 and was used sparingly until the mid-1980s when it was sold to a radio club. (Courtesy Howard Meile.)

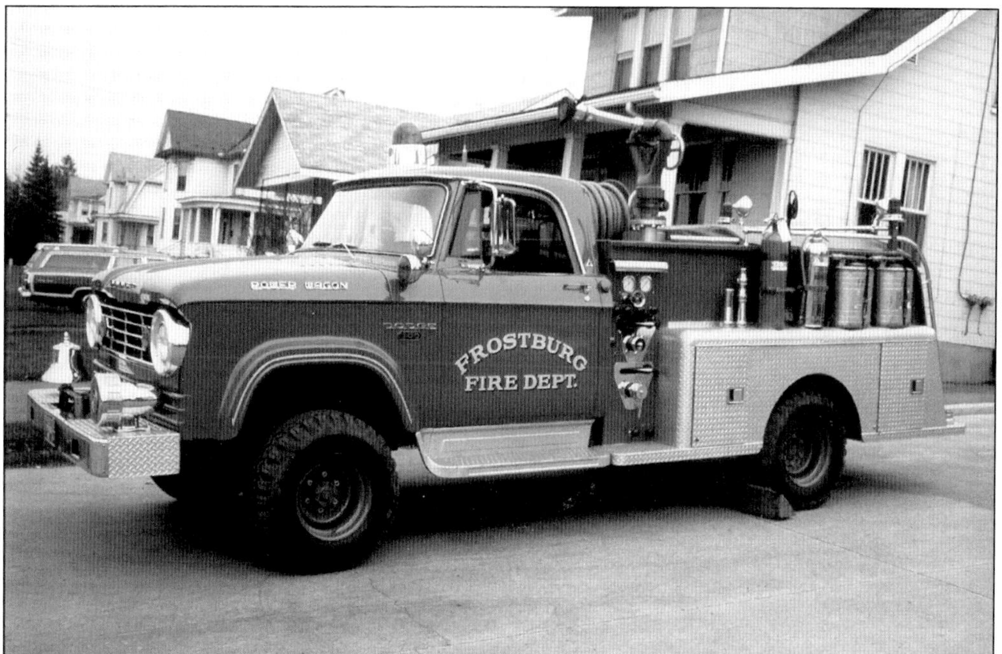

This 1967 Dodge/Bruco mini-pumper was purchased by Frostburg and had a 350-GPM pump and a 250-gallon water tank. It responded to many calls over the next 20 years, and had the body replaced in 1987. It then served as a brush truck until the mid-1990s. This apparatus was used at every type of fire from grass fires to main structure fires. Note the big turret pipe.

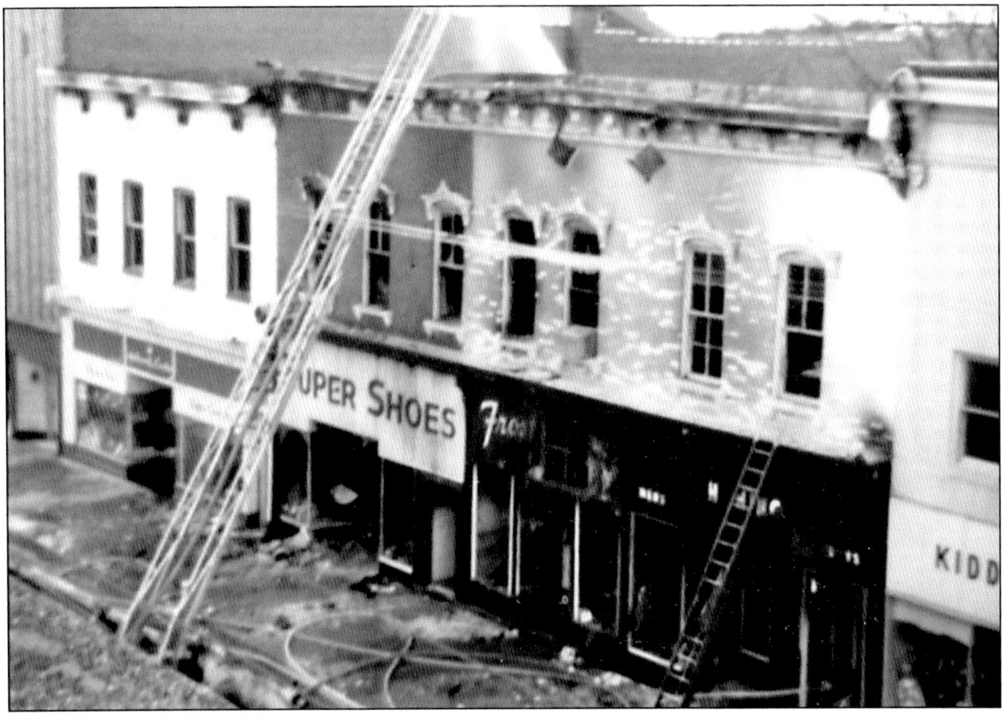

This view of Frostburg's spectacular January 1973 blaze was taken the following morning some 15 hours after the first alarm. The extended ladder with attached hose and nozzle was still pouring water on the ruins. (Courtesy Frostburg Museum.)

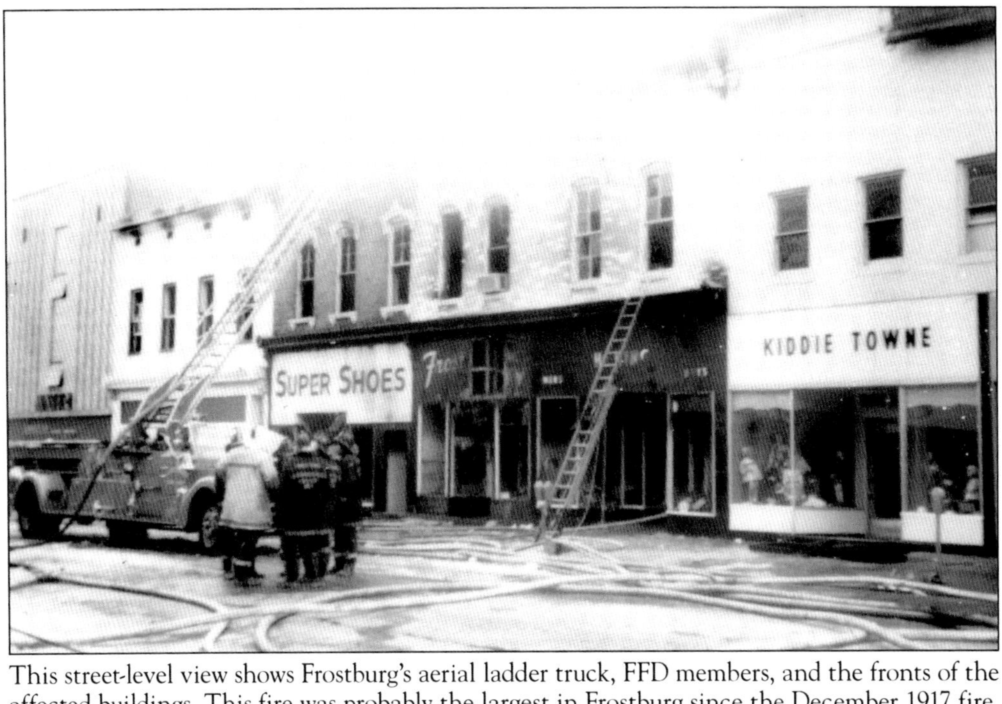

This street-level view shows Frostburg's aerial ladder truck, FFD members, and the fronts of the affected buildings. This fire was probably the largest in Frostburg since the December 1917 fire. (Courtesy Frostburg Museum.)

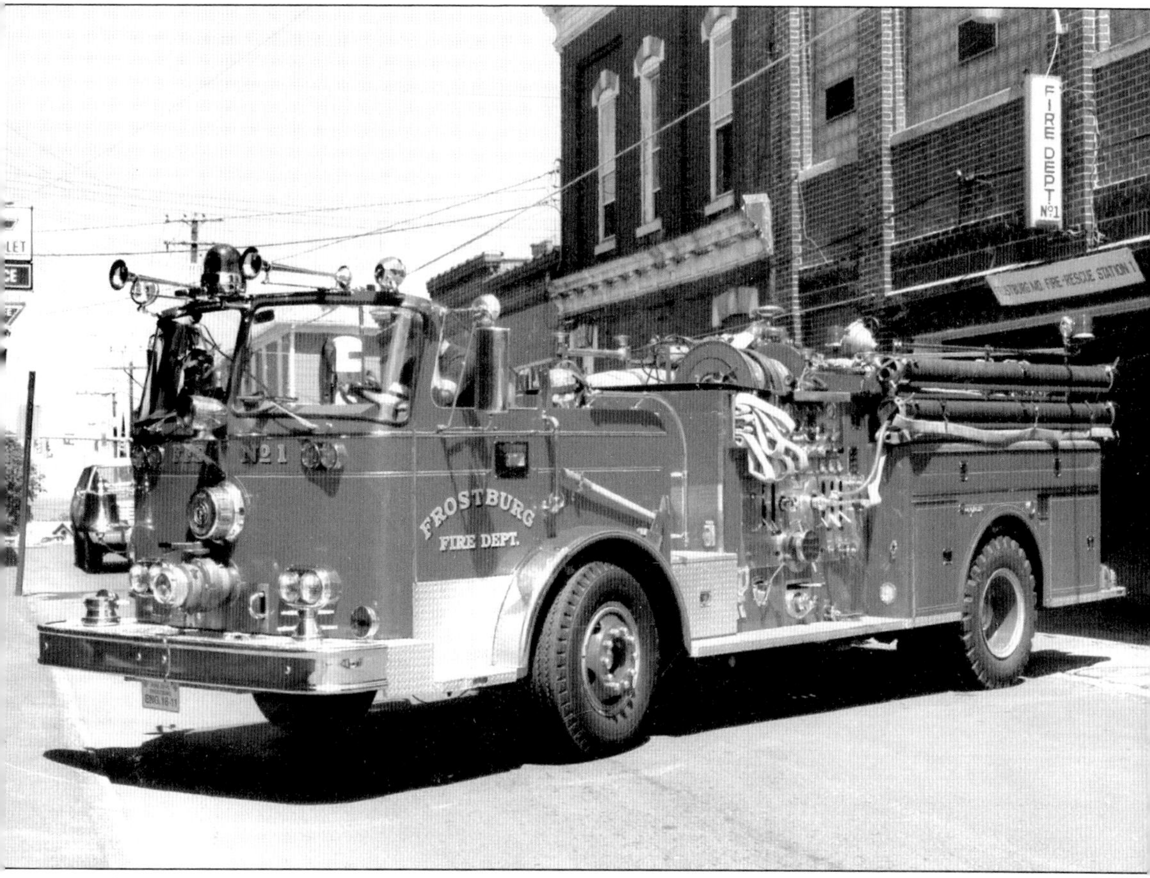

Frostburg Fire Department purchased this 1973 Seagrave PB model with a diesel engine and an increasingly rare open cab. This sleek-looking truck was rebuilt with a closed cab and new body in 1990. In 1996, it was sold to Confluence, Pennsylvania; its current whereabouts are unknown. It could pump 1,000 GPM and carry 500 gallons of water. (Courtesy Virginia McLaughlin.)

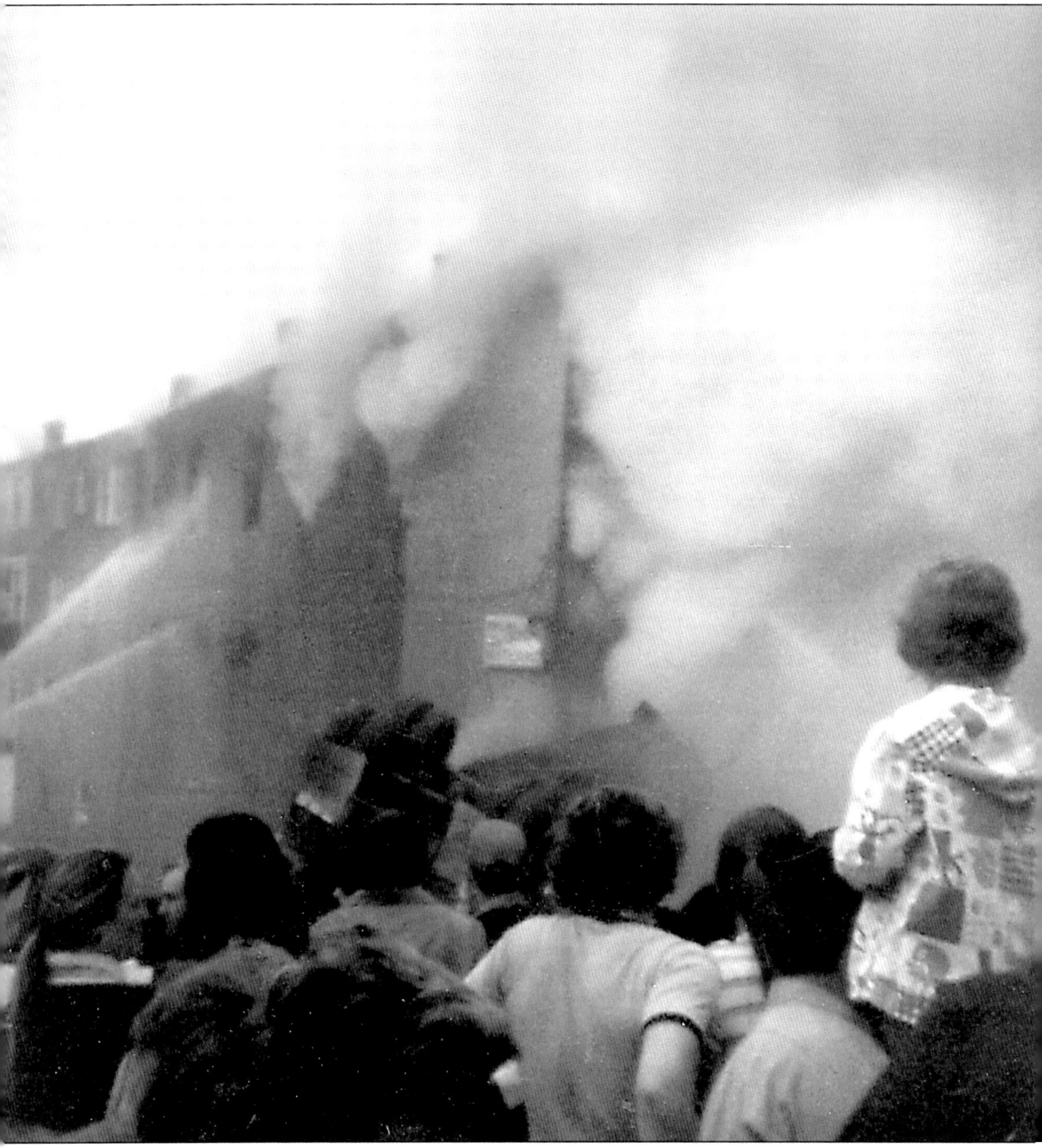

This was the view from the Frostburg Elk's Home on East Main Street as spectators stare at the upset truck trailer sticking out of Stoop's Restaurant building. The disaster happened February 18, 1981, and resulted in three deaths, many injuries, and complete destruction of the affected building.

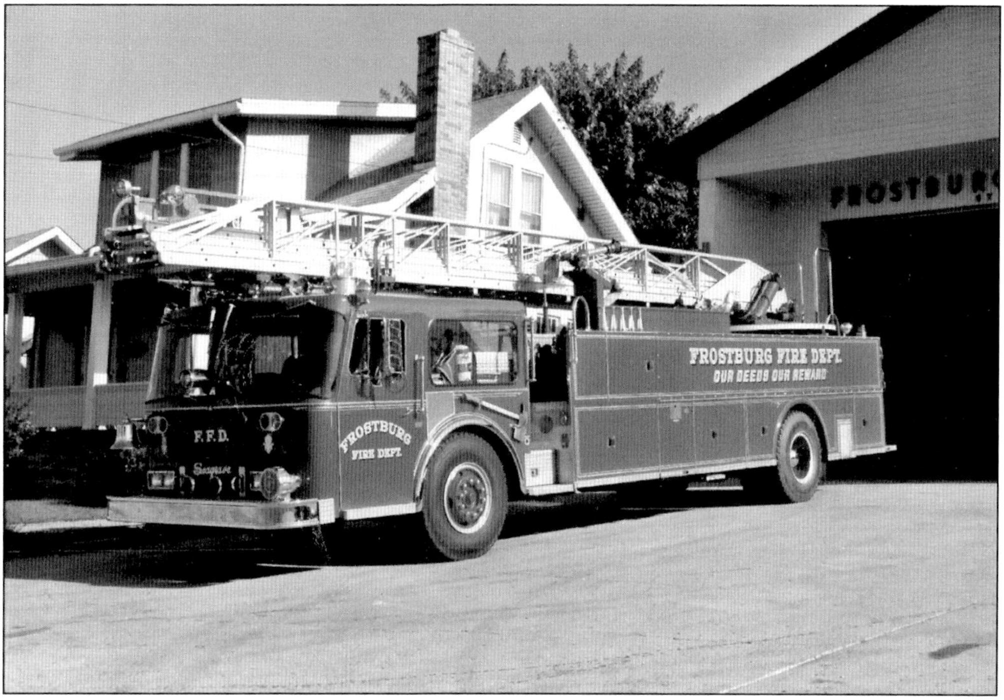

Good fortune came to FFD in December 1981 when a state grant through Frostburg State College's budget brought this 1981 Seagrave WR model 100-foot aerial ladder truck. (Courtesy Howard Meile.)

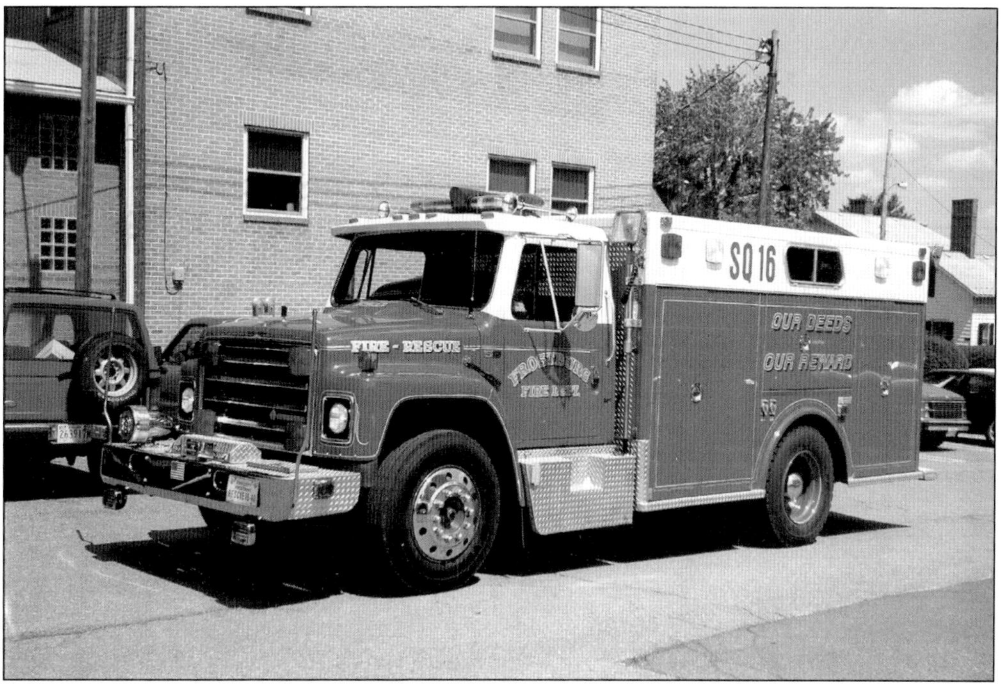

Replacing the 1958 Dodge/Bruco rescue truck was this 1986 International/Marion rescue rig. It has seen heavy use in its almost 20 years of service.

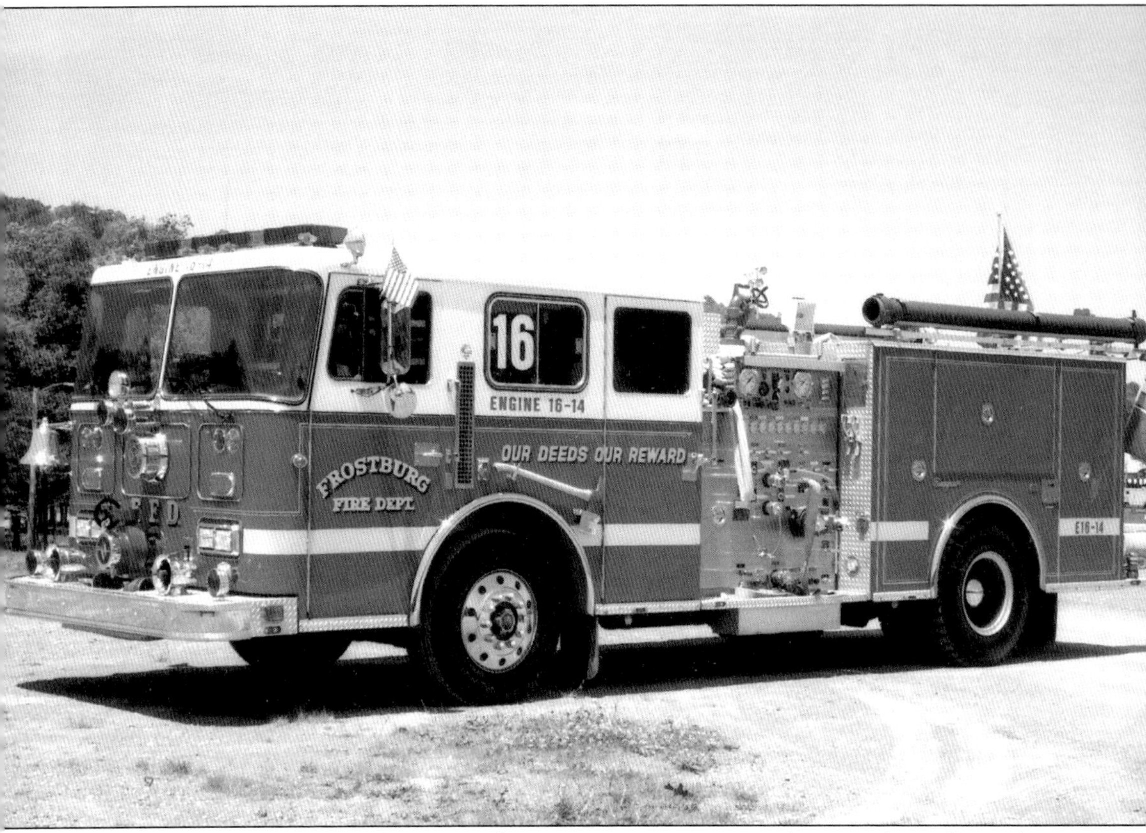

This 1988 Seagrave JB model pumper has a 1,250-GPM pump and carries 500 gallons of water. It is shown in the new white-over-red paint scheme at a parade in Bedford, Pennsylvania, in June 2004.

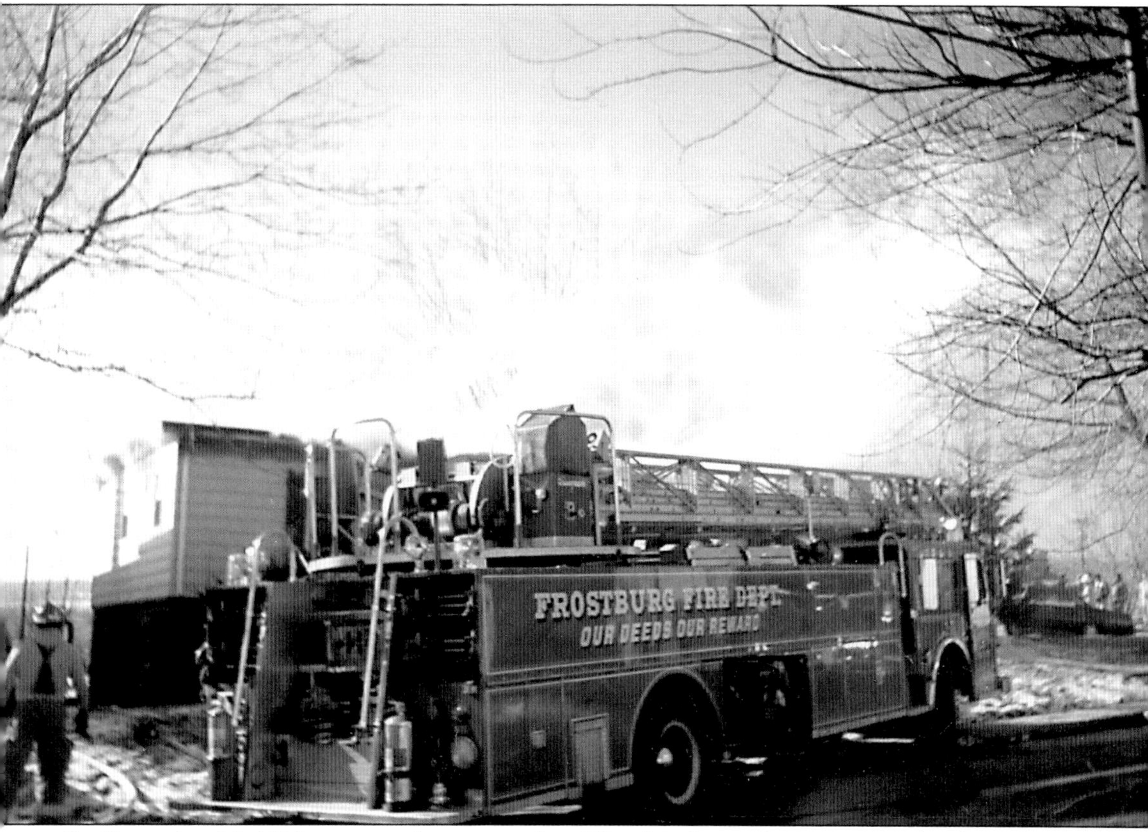

On December 25, 1990, a serious fire occurred on West Mechanic Street on top of Federal Hill. This view shows Frostburg's aerial ladder parked in front as flames begin to break through the roof of the house.

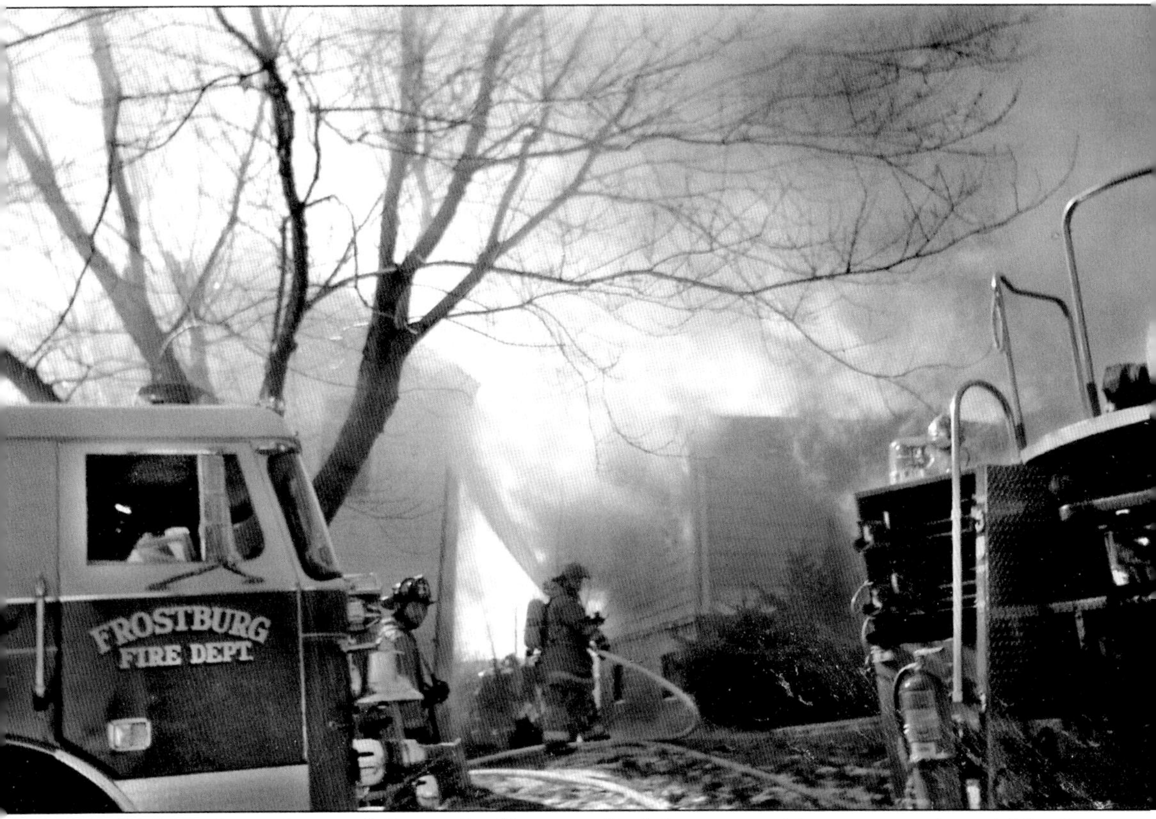

This view shows the now fully involved house as firefighters scramble to protect the neighboring house. FFD was assisted by three nearby fire companies. Many volunteers had a change of holiday plans.

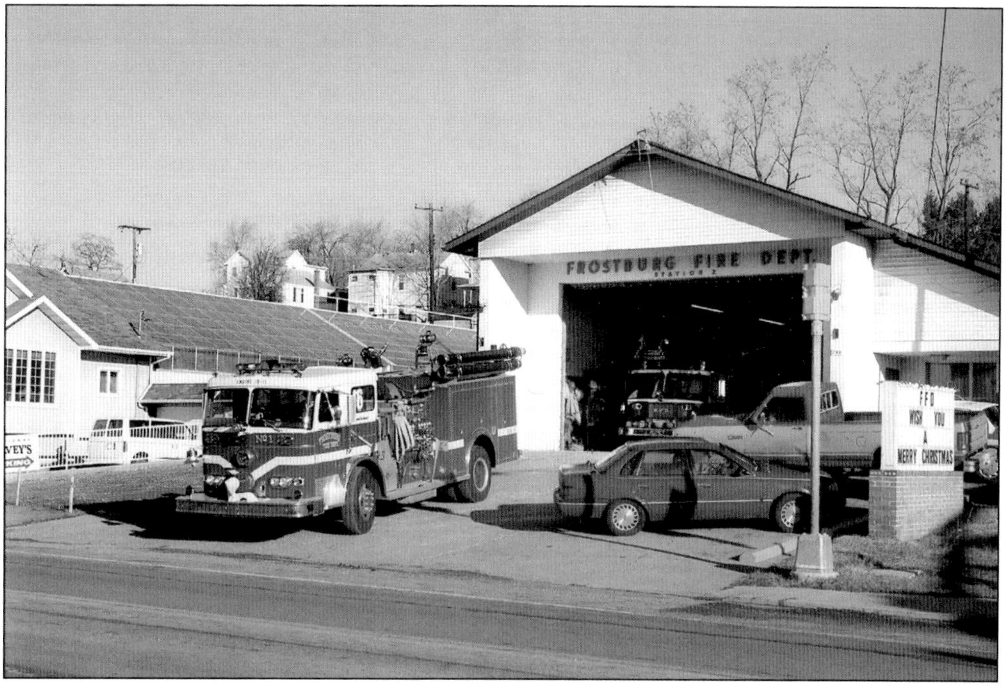

The rebuilt 1973/1990 Seagrave/Bruco pumper is seen responding to a call from Station No. 2 in the mid-1990s. By this time, Station No. 2 had received a peaked roof and traffic signals.

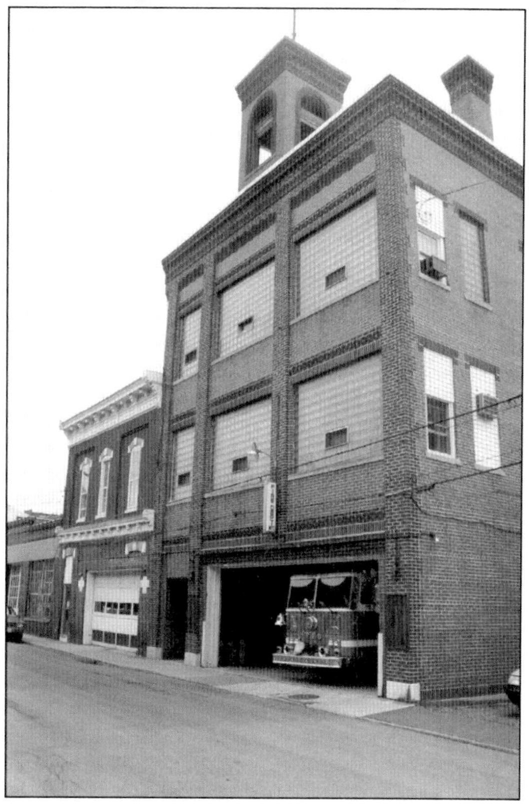

After 70 years, Station No. 1 is proving to be a problem with larger apparatus because it has a very low doorway. The 1988 Seagrave is shown at the open doorway. At this time of this photo, the Frostburg Area Ambulance occupied the old fire station/city hall next door.

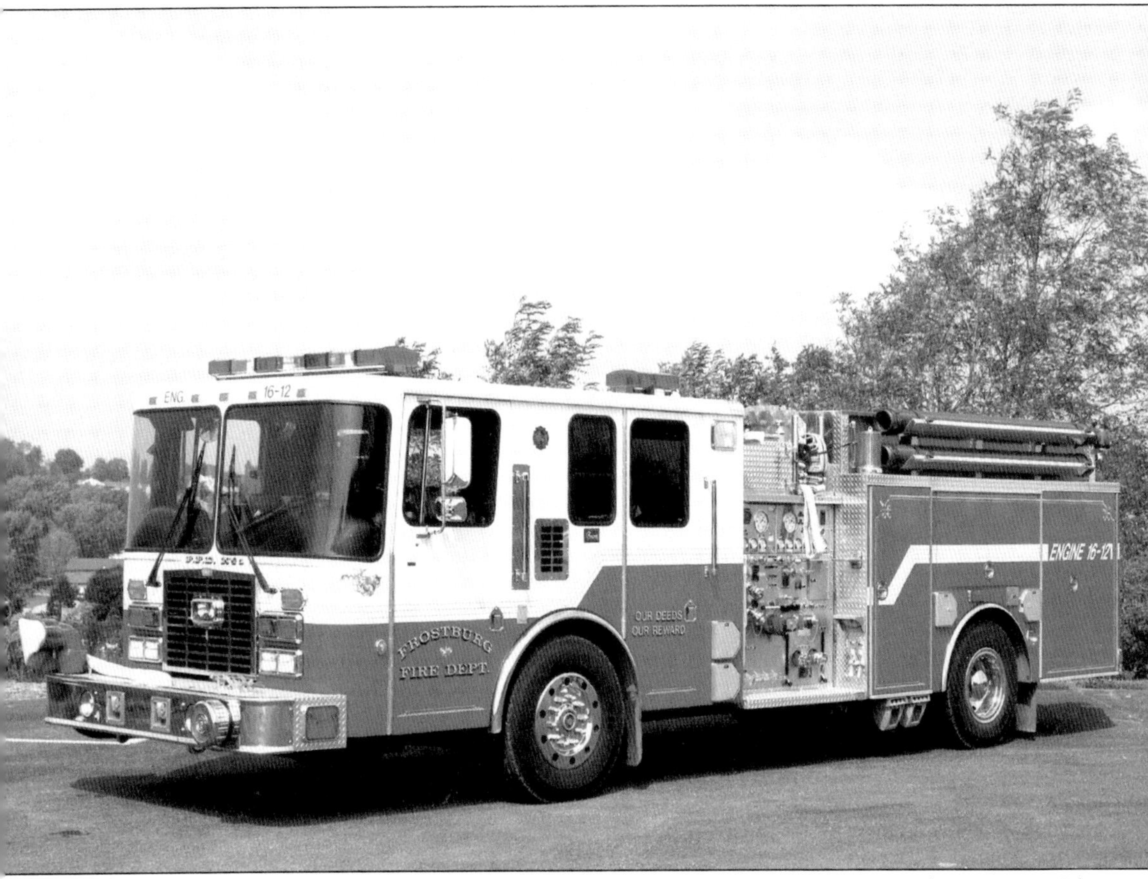

The first non-Seagrave pumper in many years for Frostburg Fire Department was this 1996 HME/4 Guys pumper built in nearby Meyersdale, Pennsylvania. The big rig has a 1,500-GPM pump and carries 750 gallons of water and 40 gallons of foam.

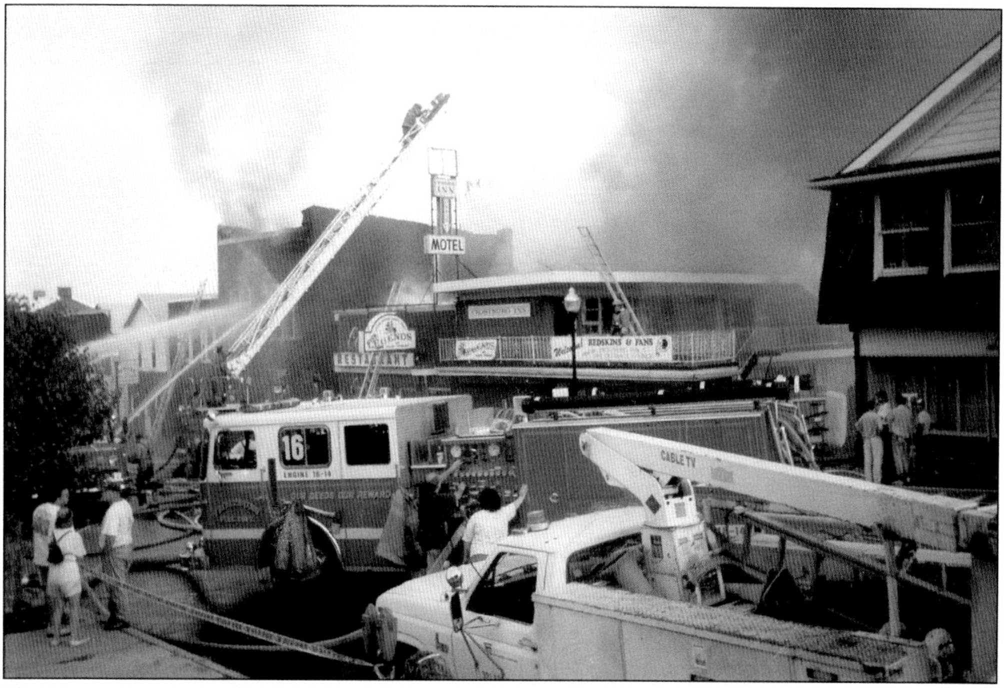

The August 20, 1996, fire is seen looking east along East Main Street as fire spreads through the affected buildings. FFD's 1988 Seagrave pumper and 1981 Seagrave ladder truck are hard at work along with mutual-aid companies and local utilities.

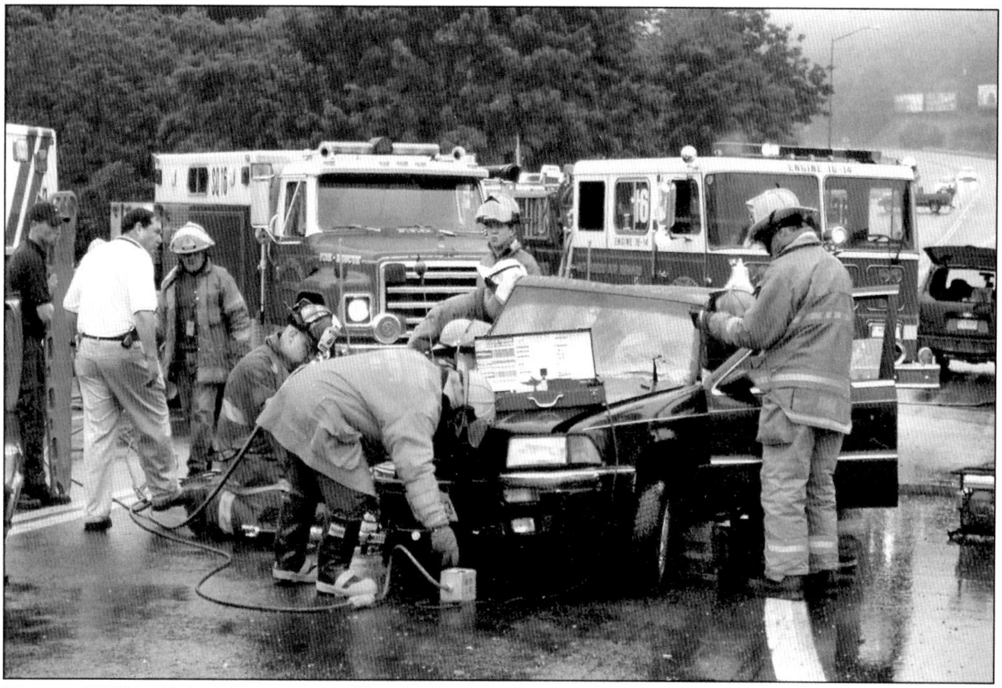

Here is FFD's 1986 International/Marion rescue squad and 1988 Seagrave pumper at a fatal highway crash in July 2003. FFD's firefighters are shown using various rescue tools to free a trapped victim.

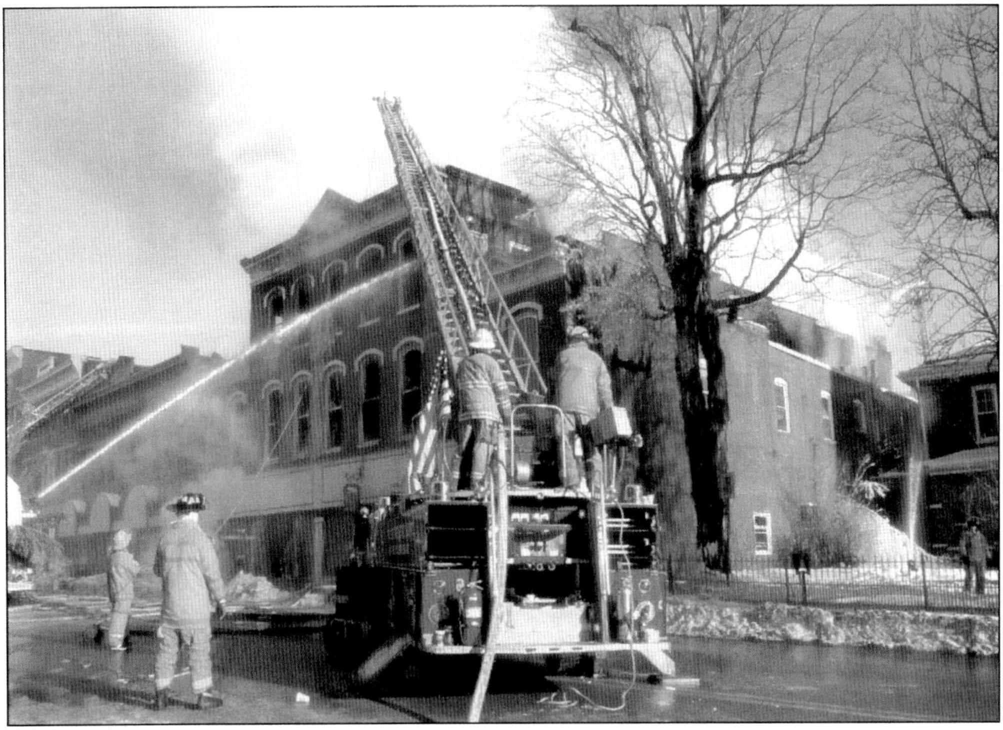

The historic Lyric Theatre building suffered this destructive blaze on February 1, 2004. Here FFD's 1981 Seagrave ladder is pouring water on the gutted structure. (Courtesy Steve Bittner, *Cumberland Times-News*.)

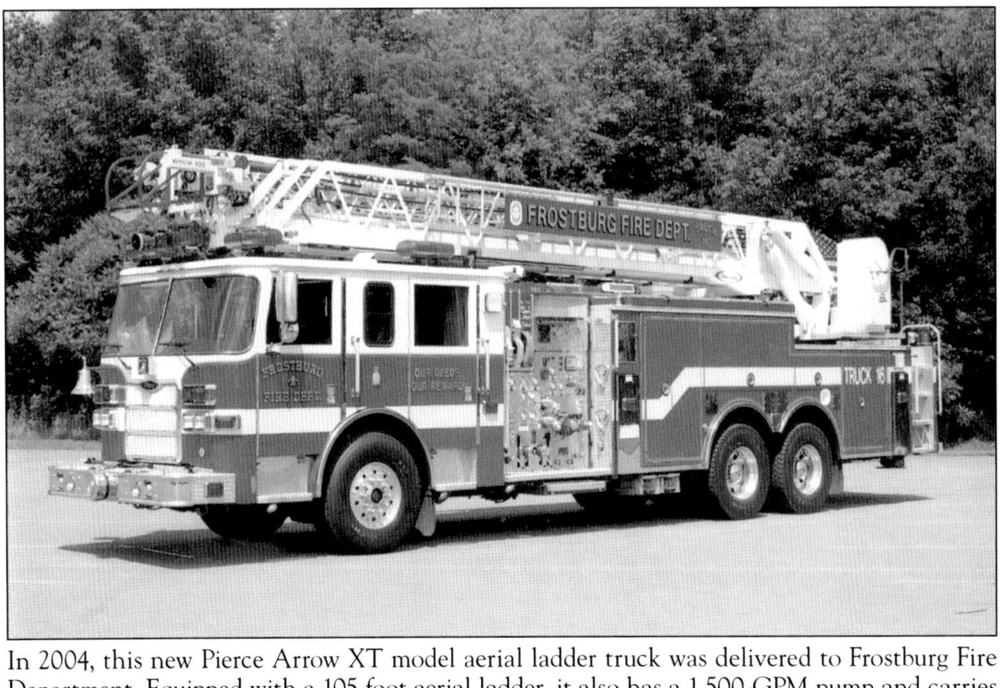

In 2004, this new Pierce Arrow XT model aerial ladder truck was delivered to Frostburg Fire Department. Equipped with a 105-foot aerial ladder, it also has a 1,500-GPM pump and carries 500 gallons of water.

Three

COAL COUNTRY FIREFIGHTING

"Coal Country" in Allegany County can be loosely defined as starting around Mount Savage in the northwest part of the county, through Frostburg and surrounding area, south along Georges Creek about 16 miles to Westernport. This area began to be exploited for its coal resources in the 1840s when railroads came to the area. Within four decades, coal was a vital part of the Allegany County economy.

After a major fire in 1876, Westernport organized its first fire protection by 1881. Over the next 30 years, the organization would change several times until Potomac Fire Company No. 2 was organized in 1912. Motorized apparatus soon replaced the "Little Giant" chemical cart and the "Little Mountaineer" and "Big Mountaineer" hand pumpers.

Lonaconing suffered a major fire in its business district in September 1881. As a result, the Lonaconing Fire Company was organized in April 1882 with a hand pumper named "Minnehaha" and several hose reels serving as the first fire apparatus. In 1883, the first station was built, and in 1906, the organization changed its name to Goodwill Fire Company No. 1.

In 1896, Mount Savage organized its first fire company. With the shops of the Cumberland and Pennsylvania Railroad in town, the shop whistle was used to summon volunteers.

Midland organized its first fire company in 1900 after several disastrous fires had occurred. The first equipment was purchased from Hagerstown, Maryland. This was a four-wheeled hose reel, and the new company also obtained a hand pumper named "Eureka."

As the paper mill and the adjacent town of Luke continued to grow, a fire company was formed in 1917. The first motorized apparatus was apparently a mid-1930s small pumper.

The town of Barton suffered a devastating fire in February 1919. Barton Hose Company No. 1 was formed less than three weeks later and obtained a hose reel as its first equipment.

In 1938, the coal mining village of Shaft formed Shaft Volunteer Fire Company. A donated hose reel was the first equipment, and a small firehouse was built.

The National Road hamlet of Clarysville formed a fire company in March 1942 and received a donated hose reel from Frostburg Fire Department No. 1. In April 1946, an army surplus International pumper was purchased.

The communal nature of coal mining and railroading often led to these industries providing the majority of volunteer membership, which also promoted teamwork.

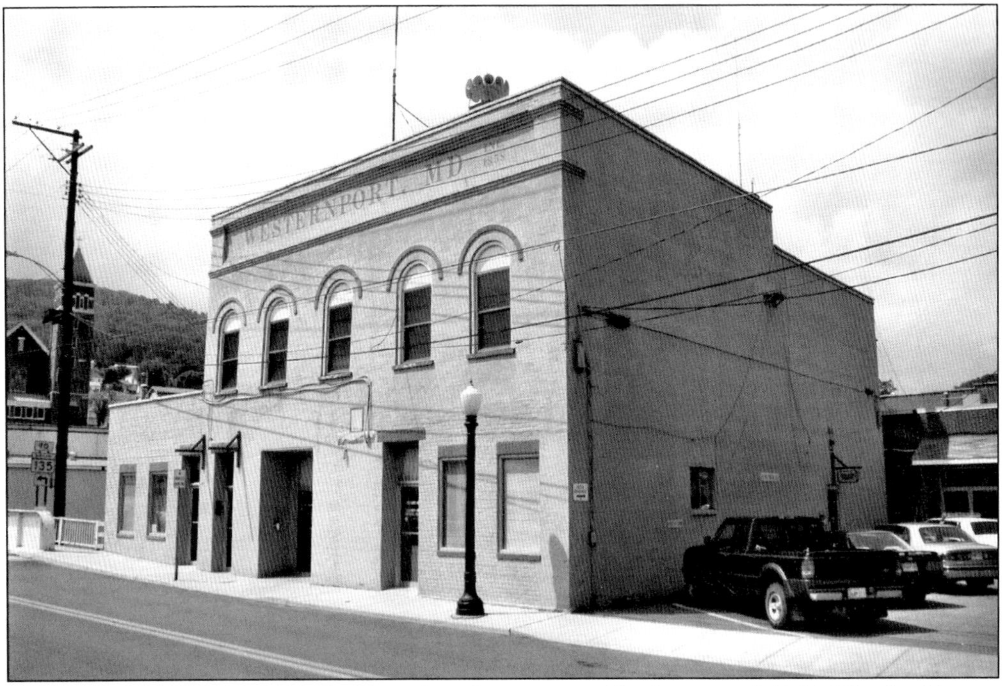

The Potomac Fire Company was housed in the municipal building until 1966. This building is located behind Potomac's current station and now houses city hall and the Westernport police.

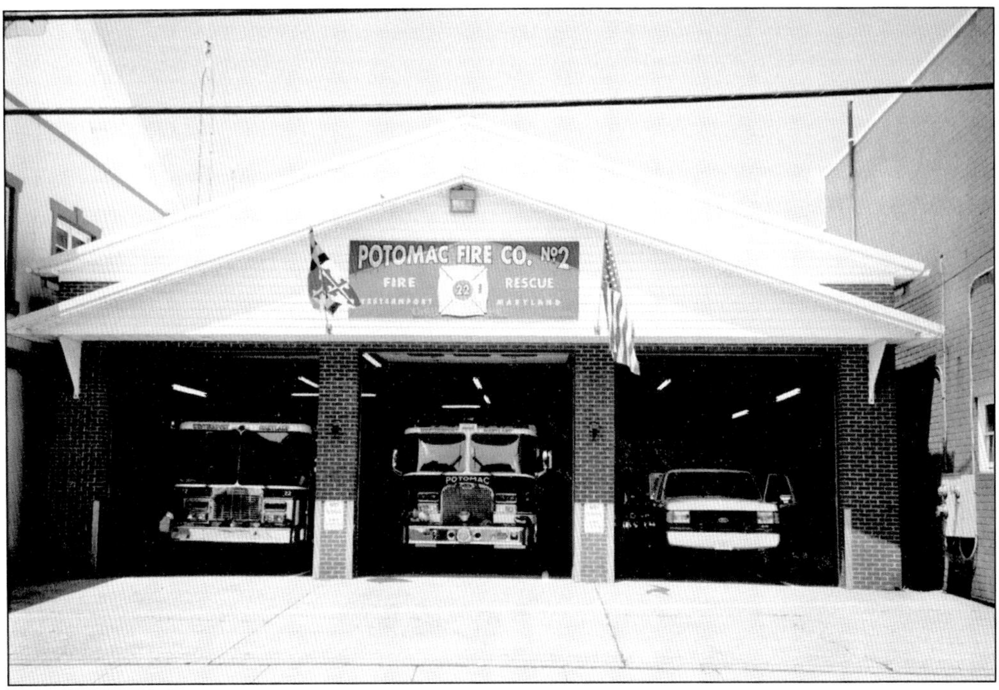

Potomac's current station was constructed in 1966 and recently had a new roof added. The apparatus visible are, from left to right, a 1991 HME/Grumman engine, a 2002 KME engine-tanker, and the 1988 Ford brush truck.

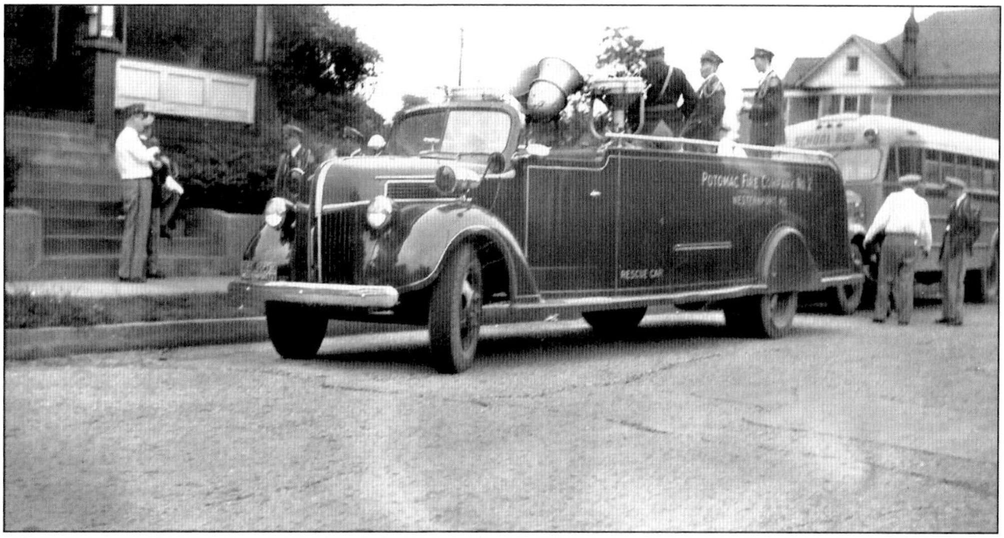

This 1941 Ford rescue car was built by an unknown builder, and the body would eventually be used to build McCoole's 1964 tank truck. Shown in 1951 during a parade, this was probably Potomac's second green-painted apparatus. A 1933 American LaFrance pumper was delivered in the now-traditional green color of Potomac Fire Company. (Courtesy William Snyder.)

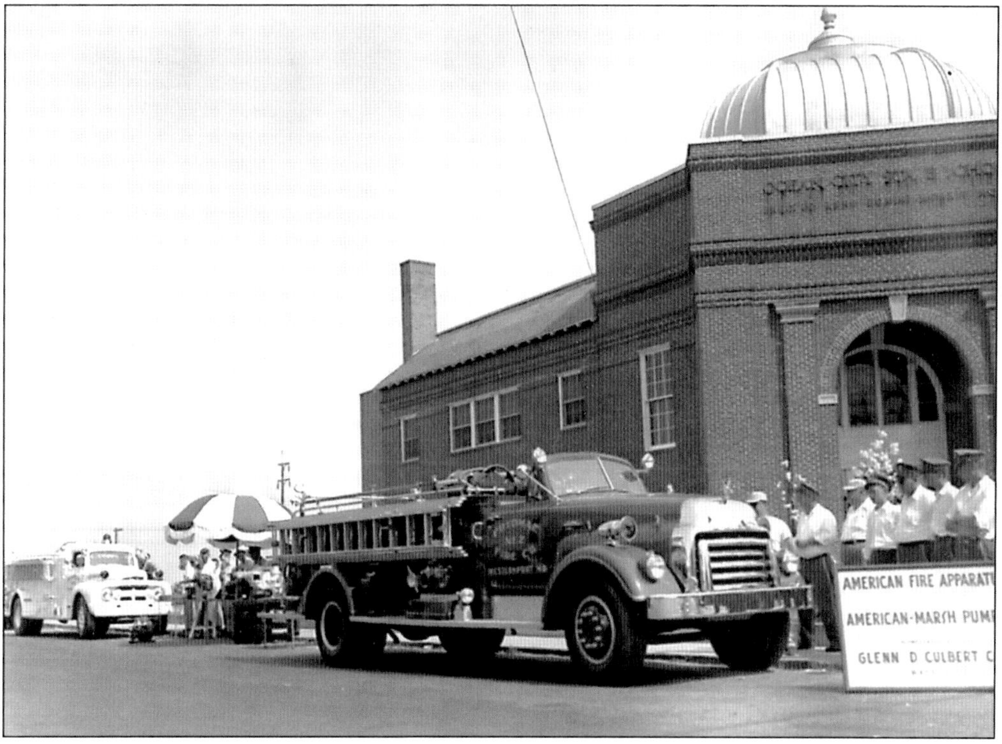

In 1953, Potomac Fire Company purchased a 1952 GMC/American pumper, which stayed in service until about 1987. It had a 750-GPM pump and originally carried 500 gallons of water. The water tank was enlarged to 1,000 gallons in the early 1980s. This photo shows the new truck on display at the 1953 Maryland State Firemen's Convention in Ocean City, Maryland. (Courtesy Ed Bosanko, Maryland State Firemen's Association [MSFA].)

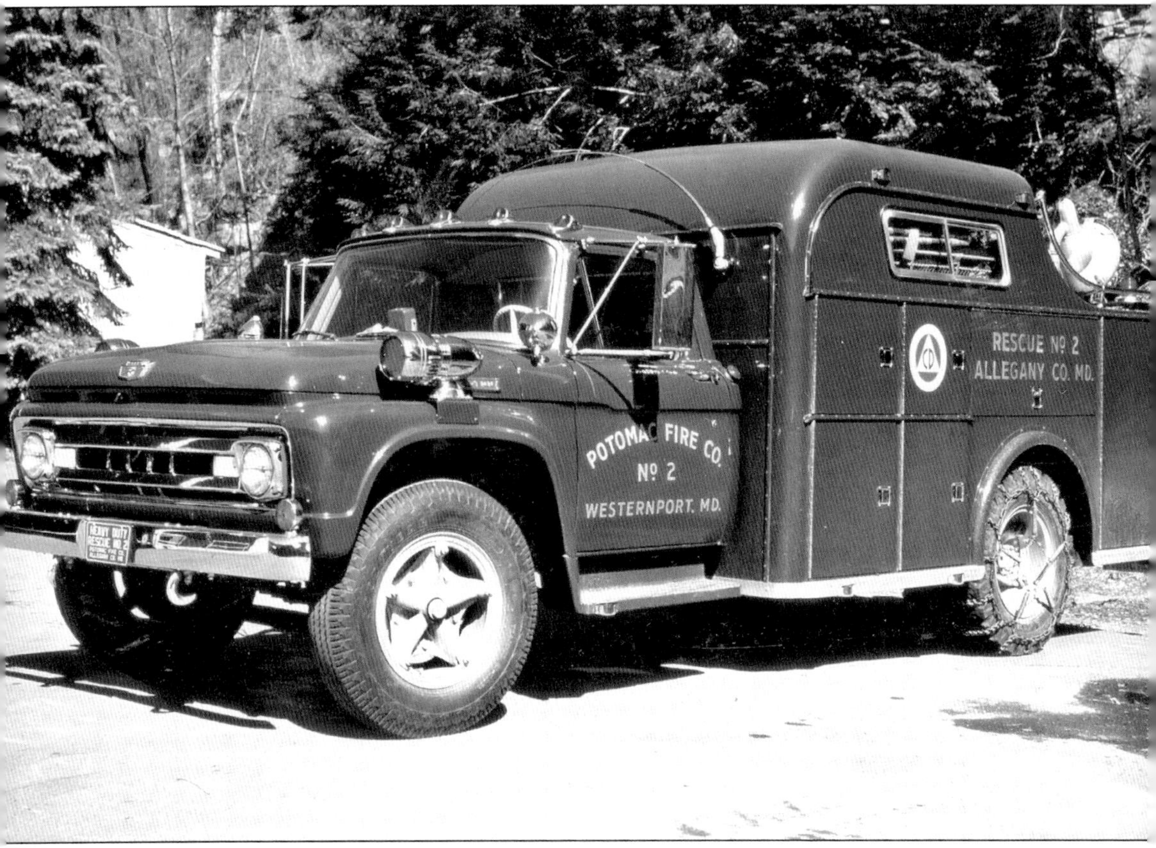

The Potomac Fire Company purchased this 1962 Ford/Gerstenslager rescue truck. It served until 1987 when it was sold to neighboring Barton Hose Company, where it lasted until the mid-1990s.

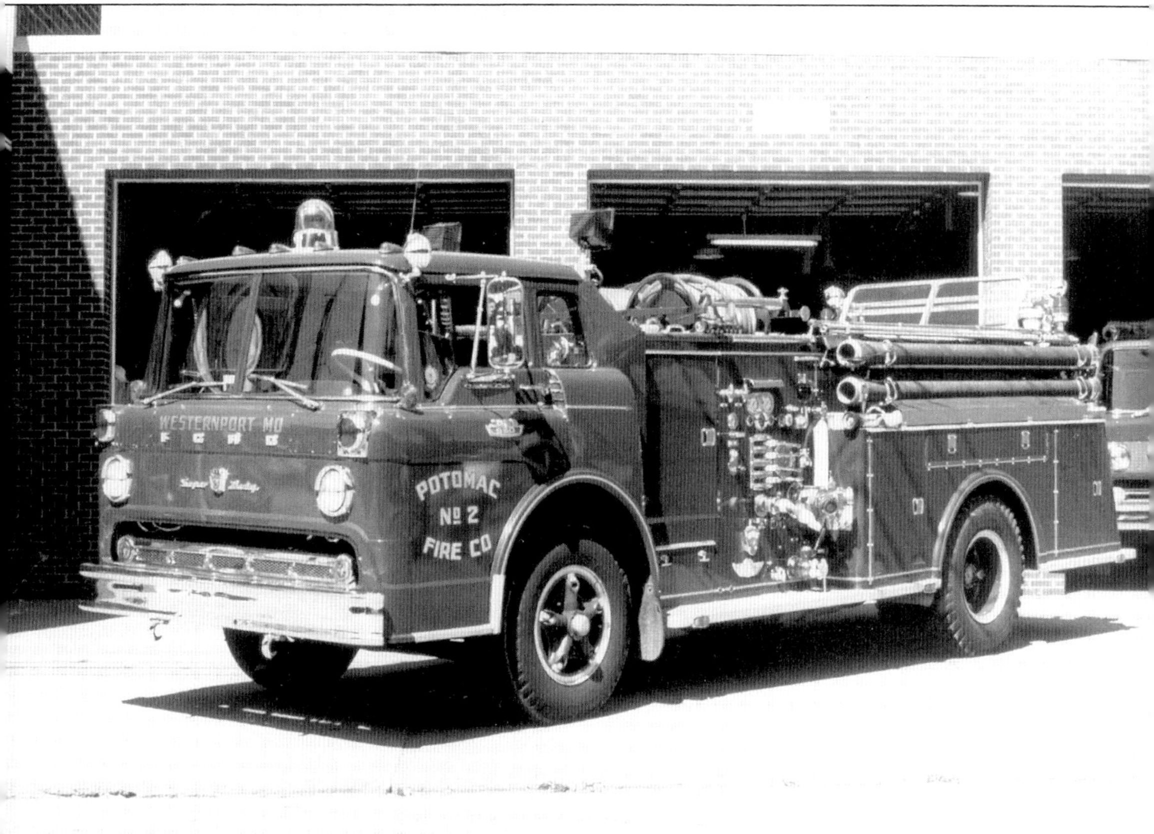

The Potomac Fire Company in 1967 purchased this Ford C/American 750-GPM pumper with a 500-gallon water tank. It stayed in service until 1991.

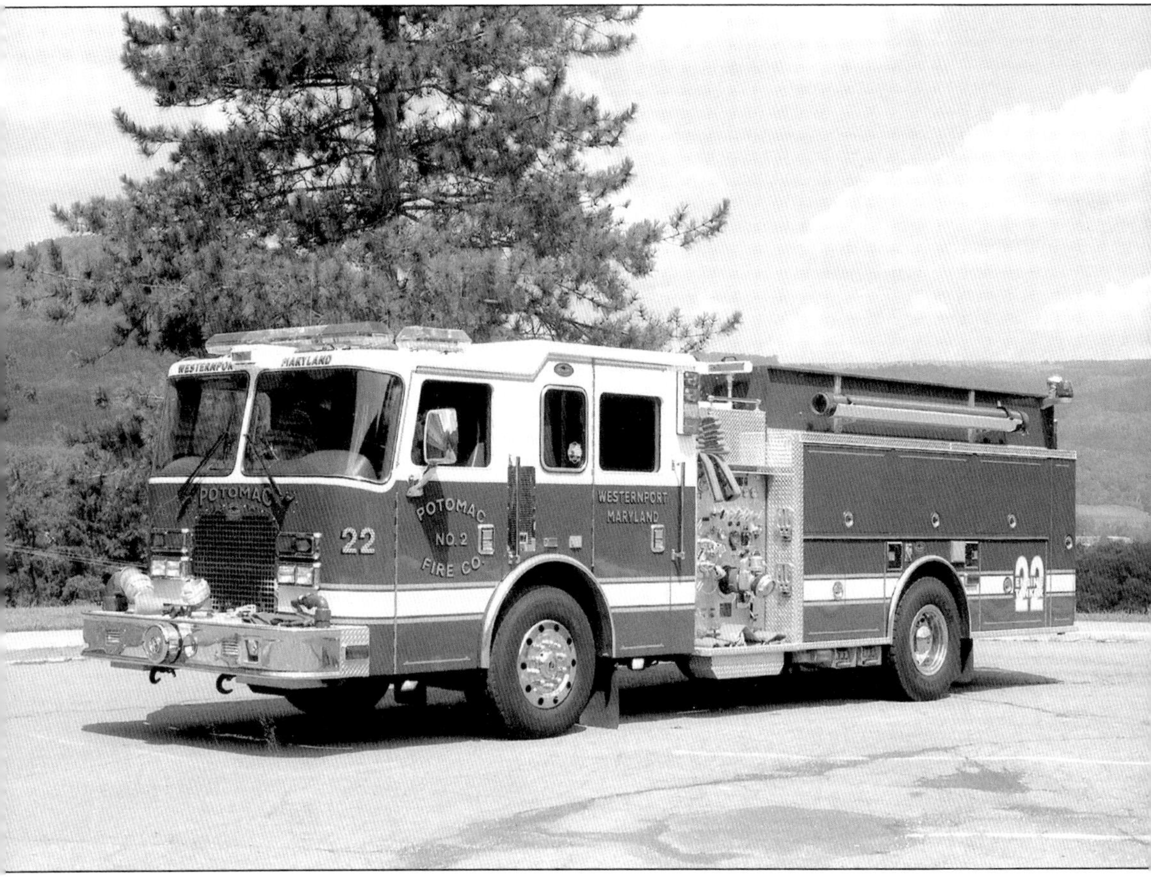

Here is Potomac's 2002 KME XL model posed at the former Bruce High School. It can pump 1,250 GPM and carry 1,500 gallons of water.

This classic firehouse served Goodwill Fire Company from 1883 until 1964 when the current station was built. The tall tower at the rear of the building was used to hang wet hose to dry. The alert siren was located on top of the tower. (Courtesy Ed Bosanko, MSFA.)

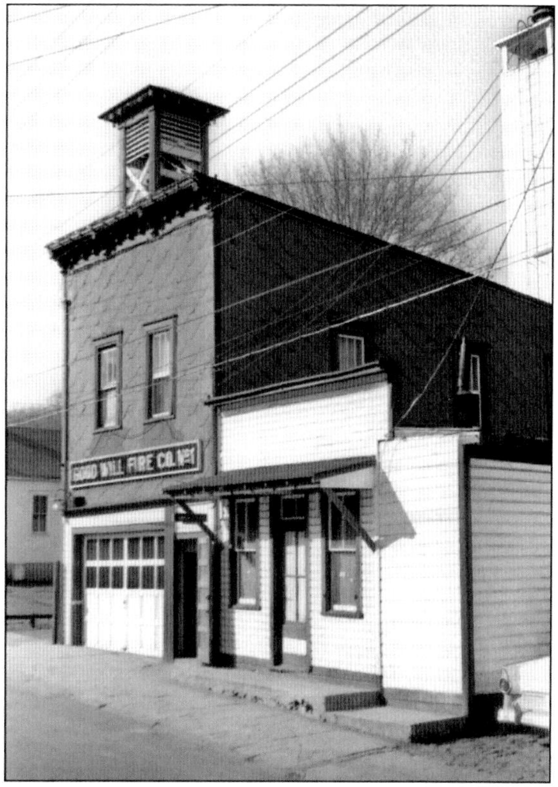

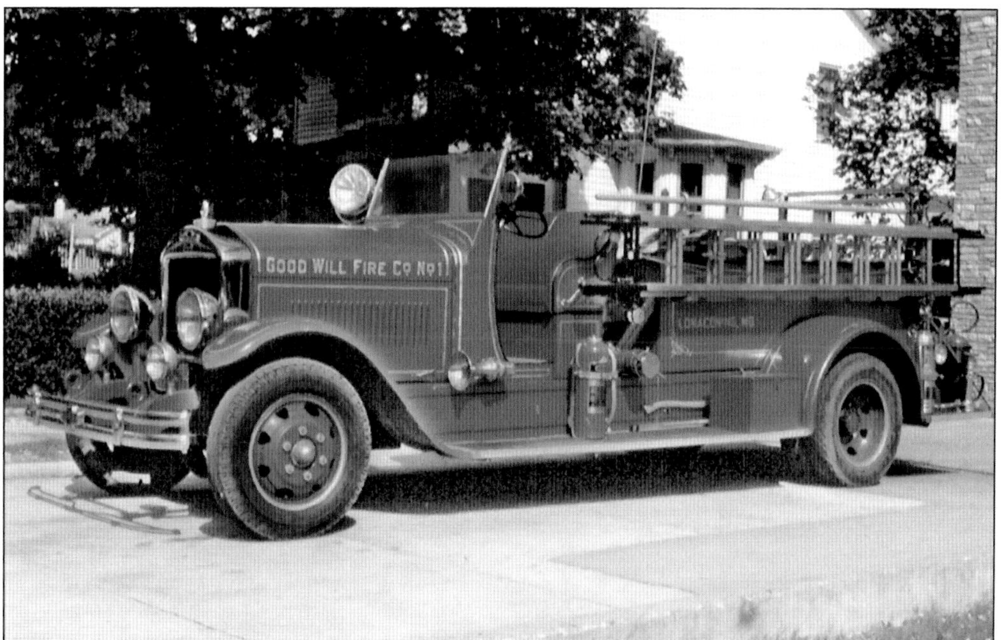

This 1973 photo shows Goodwill's 1936 American LaFrance pumper equipped with a 500-GPM pump and a 200-gallon water tank. This pumper replaced a 1921 Dodge pumper that was Goodwill's first motorized apparatus. The truck was refurbished in 1992 by the Goodwill members.

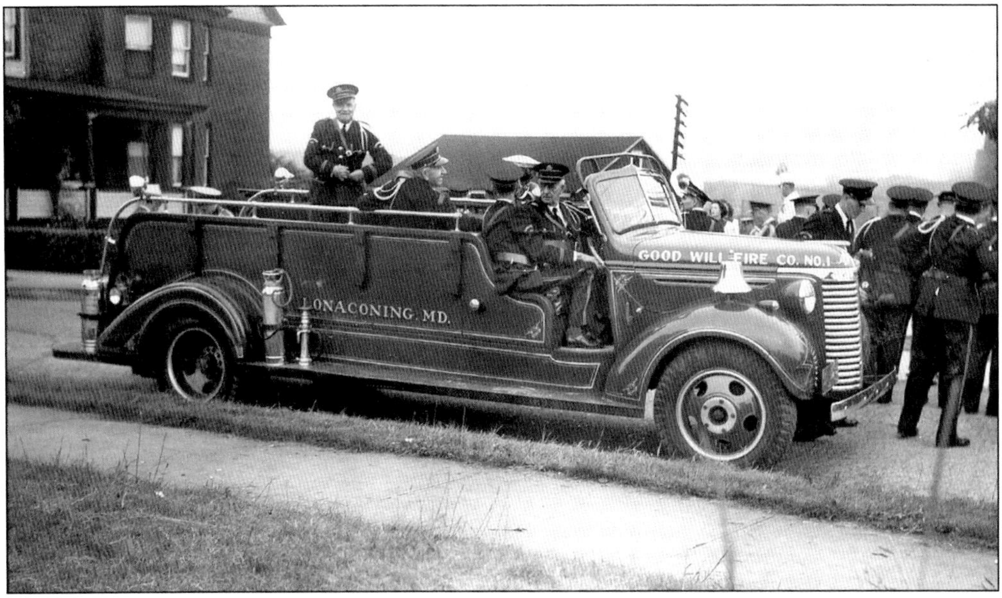

Around 1950, Goodwill firemen in full dress uniform are shown preparing for a parade. The squad truck is a 1940 Chevrolet/Brumbaugh Body Works, better known as "Bruco," of Altoona, Pennsylvania. (Courtesy William Snyder.)

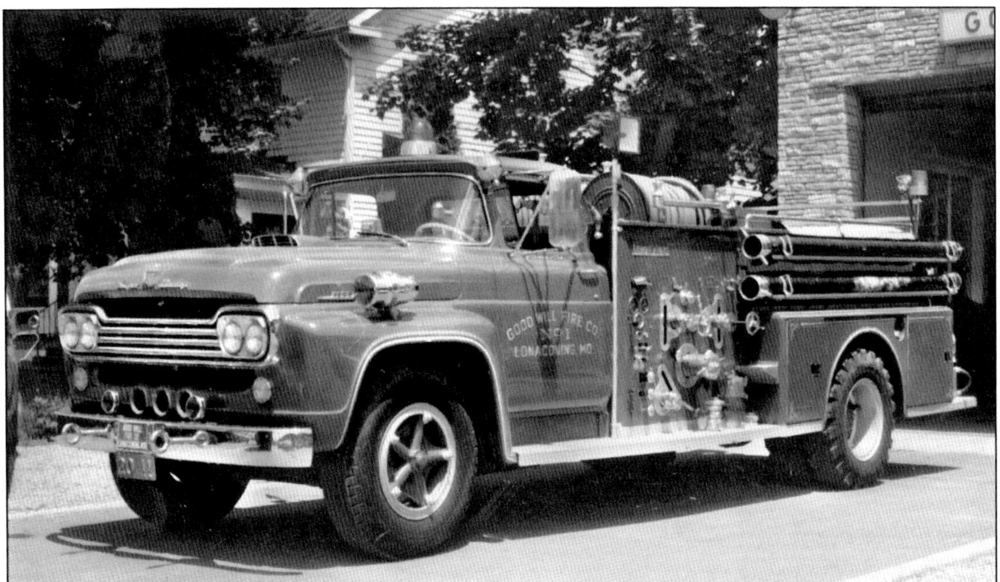

This handsome 1960 Ford/American LaFrance pumper had a 750-GPM pump and a 500-gallon water tank. In May 1990, a major fire on Main Street in Lonaconing caused scorched paint, melted warning lights, and other damage that resulted in the truck being removed from service. (Courtesy Virginia McLaughlin.)

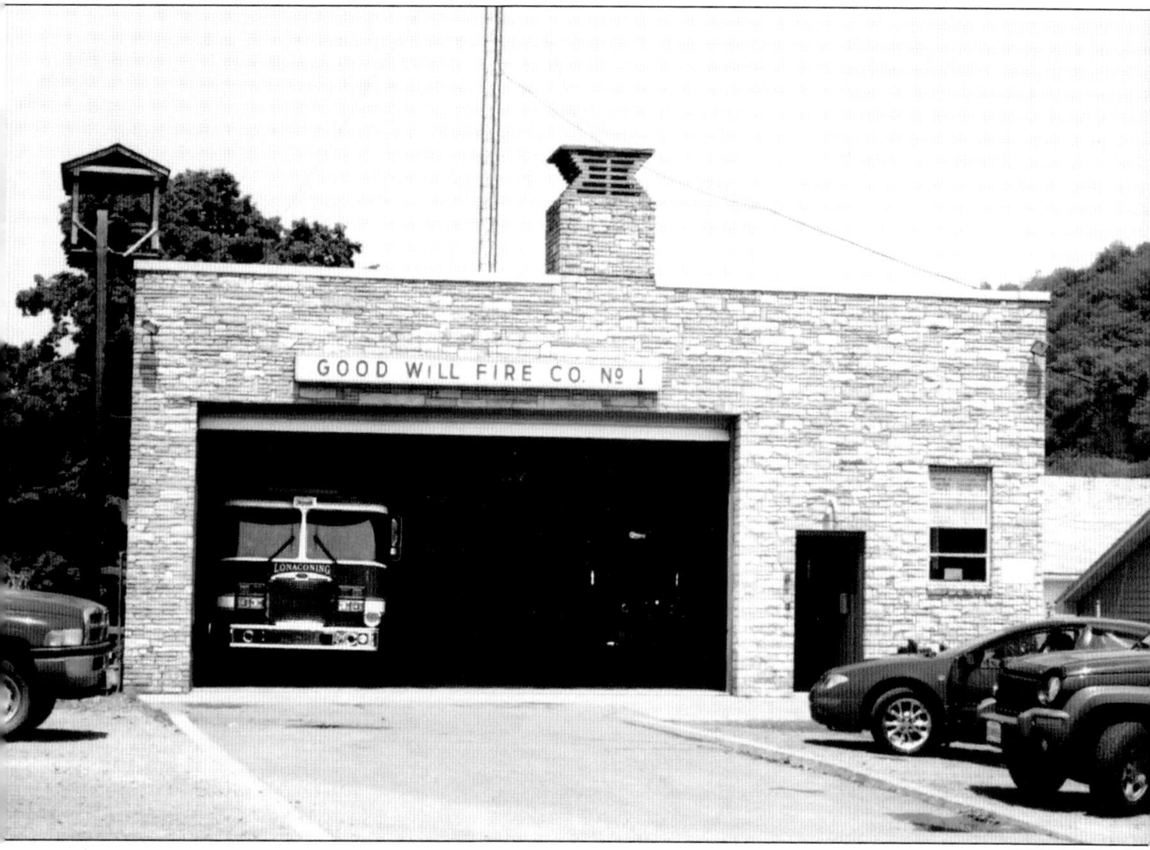

The new fire station was constructed in 1964 at the rear of the old station's property. The old station was then razed and the space used for parking. The present station now has the east wall (right side of the photo) painted as a U.S. flag.

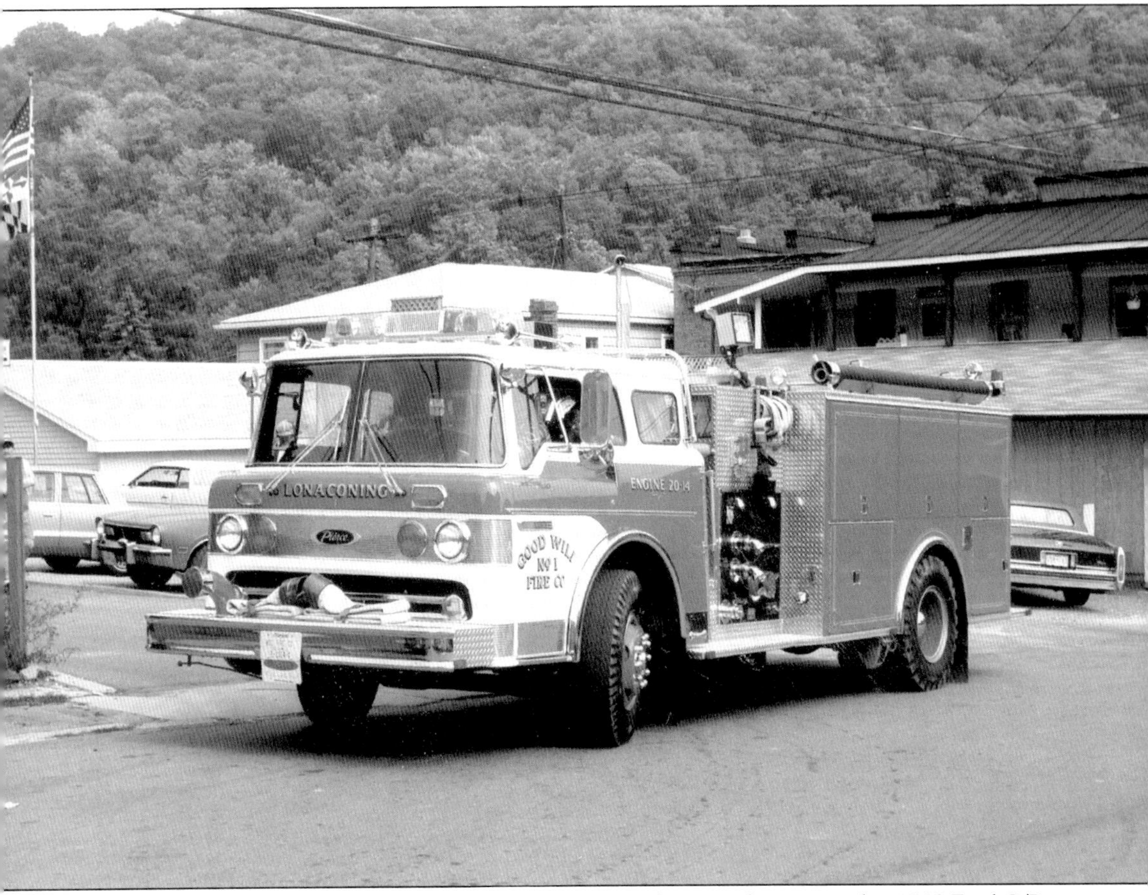

The first Goodwill apparatus with a new white-over-red paint scheme was this 1983 Ford C/Pierce pumper. It had a 1,000-GPM pump and a 1,000-gallon water tank. (Courtesy David Bowen.)

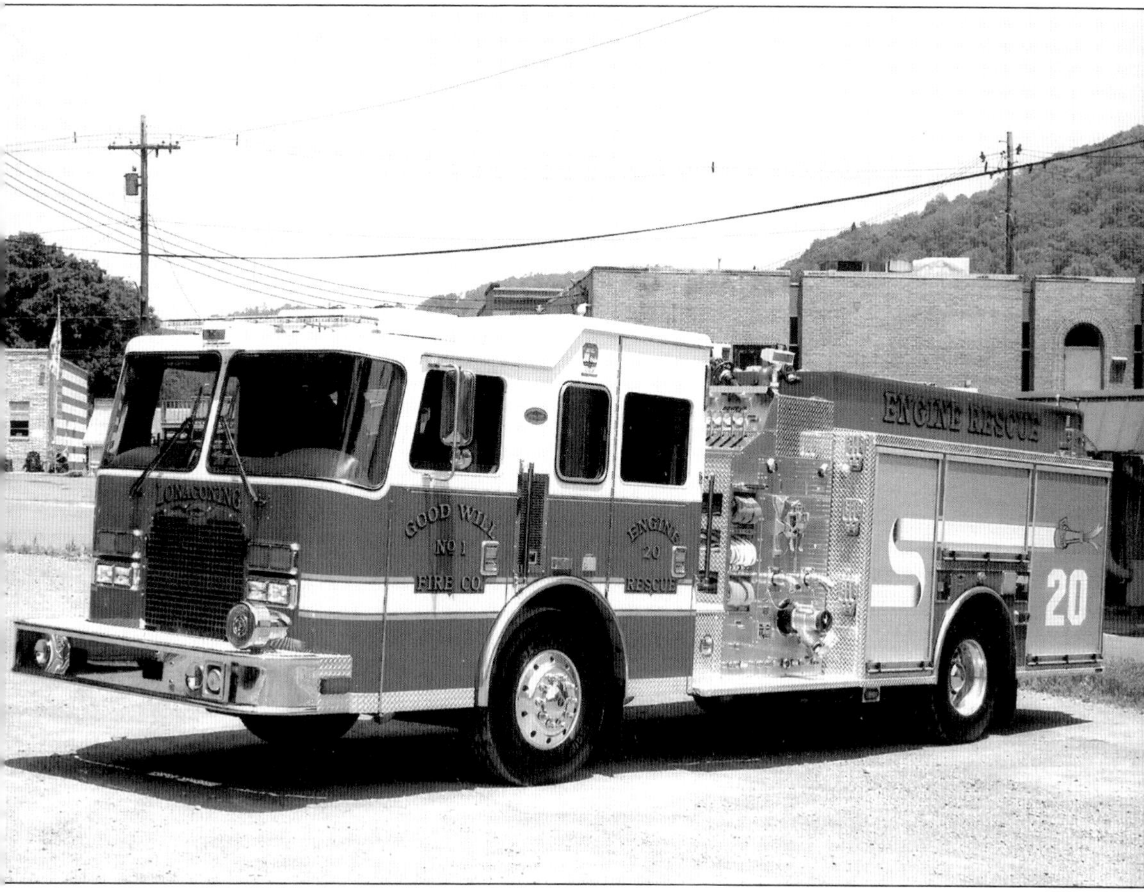

In 2004, Goodwill Fire Company bought this KME Excel model rescue pumper. It can pump 1,250 GPM and carry 750 gallons of water and a full allotment of rescue tools. The 1983 Ford/Pierce pumper was then sold to a department in Missouri.

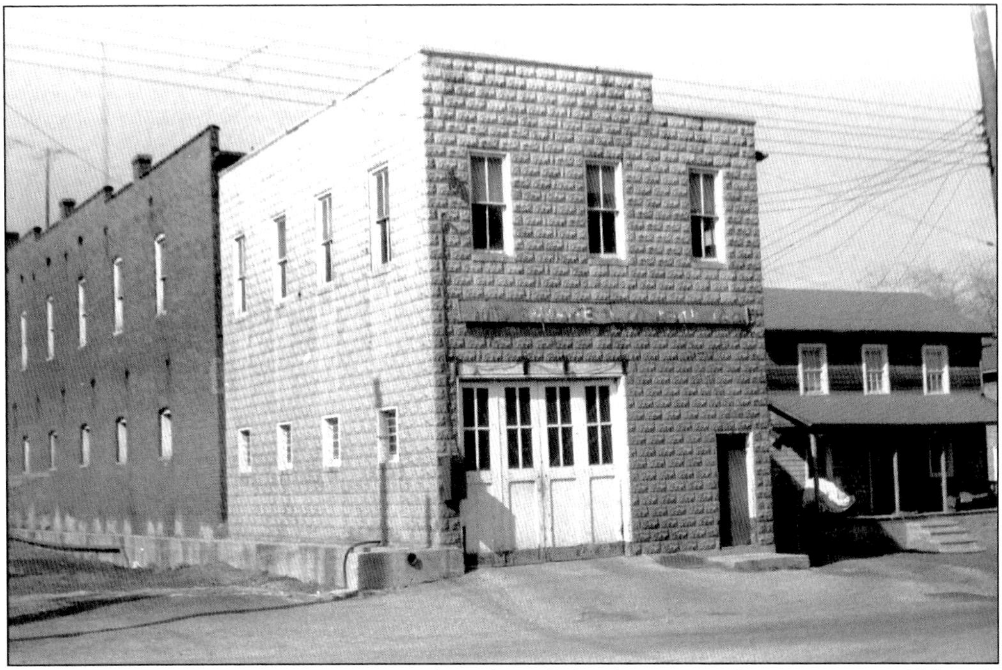

Mount Savage Volunteer Fire Company (VFC) used this fire station for many years. Although the building appears large, only the front concrete block section is the actual station. In later years, four pieces of fire and rescue apparatus would be squeezed into this building. (Courtesy Ed Bosanko, MSFA.)

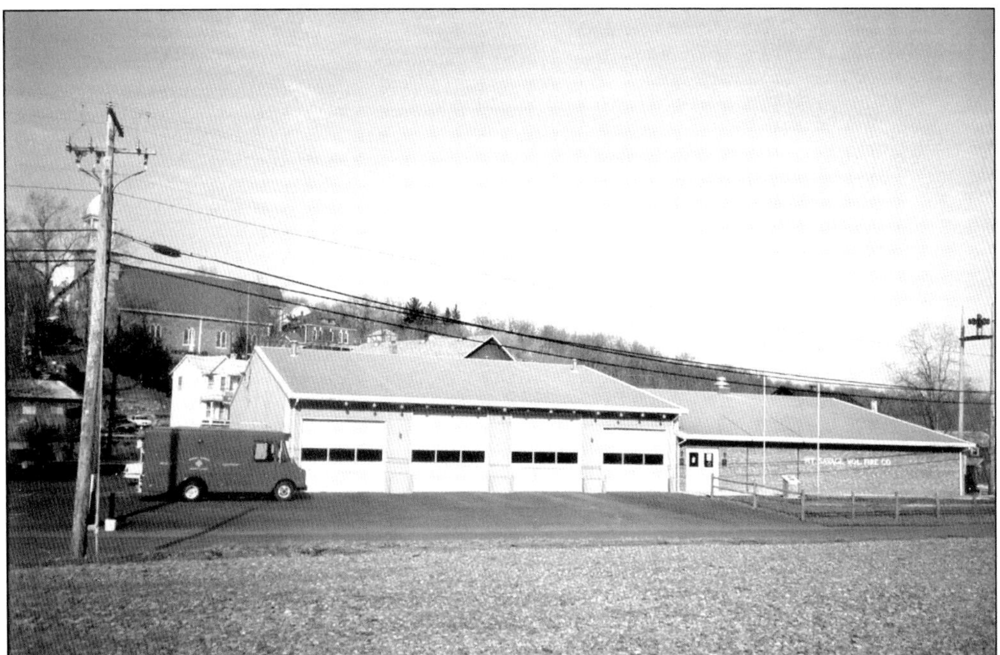

In the mid-1980s, Mount Savage VFC raised funds to build this modern fire station just down the street from the old station. The step-van shown served as the fire company's rescue truck for almost 20 years from 1978 to 1997.

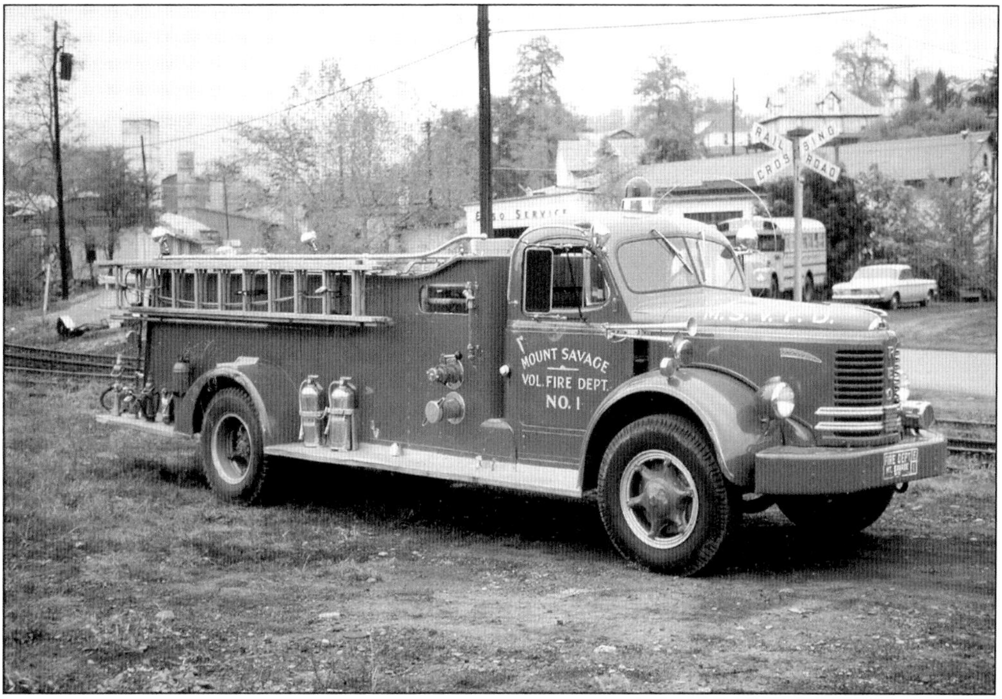

This 1951 REO/Oren pumper replaced a wartime Ford pumper acquired in 1944. The new pumper could pump 500 GPM and likely carried at least 500 gallons of water.

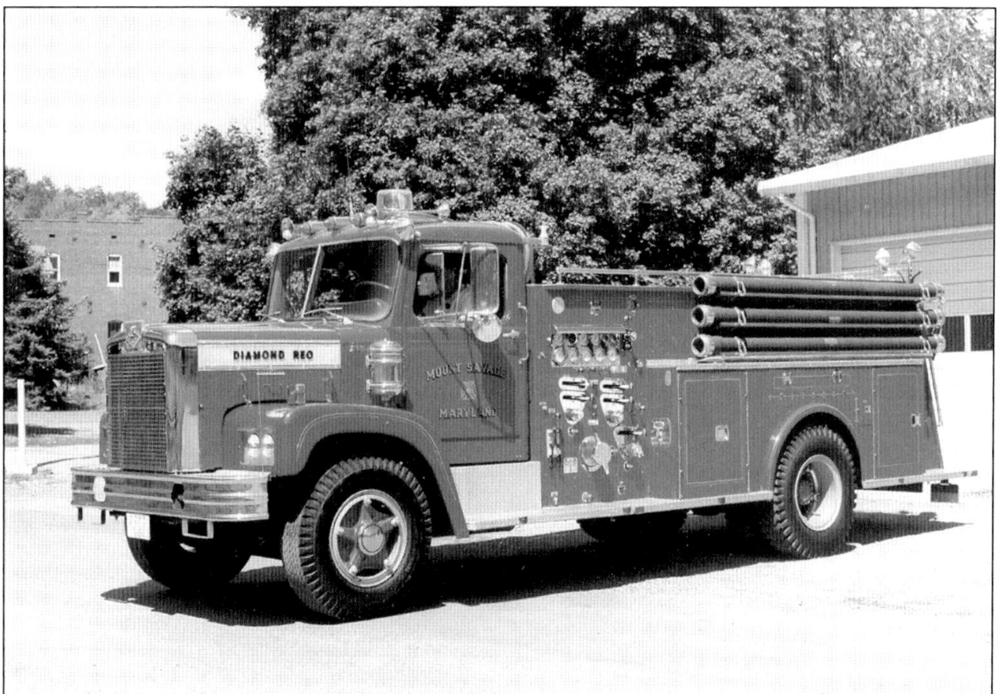

A rare delivery to this area was this 1971 Diamond Reo/Boyer pumper, which had a 750-GPM pump and a 750-gallon water tank. (Courtesy Virginia McLaughlin.)

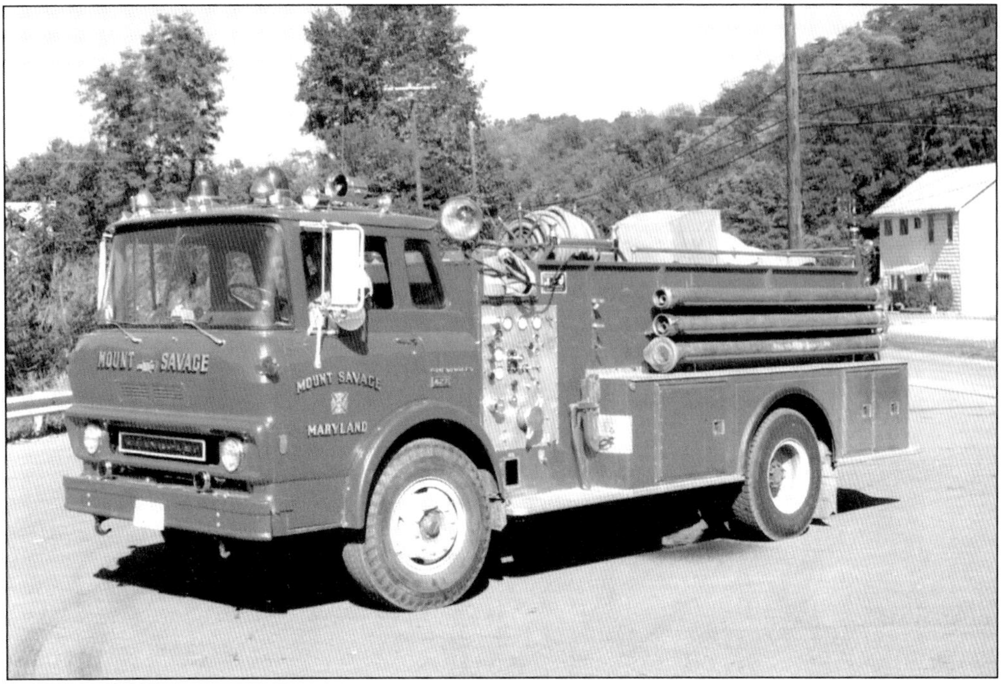

This 1976 Chevrolet tanker truck was built in nearby Somerset, Pennsylvania, by a dump truck manufacturer named J and J Bodies. This outfit built only a few fire and rescue trucks. This tanker had a 500-GPM pump and a 1,500-gallon water tank. (Courtesy John Floyd.)

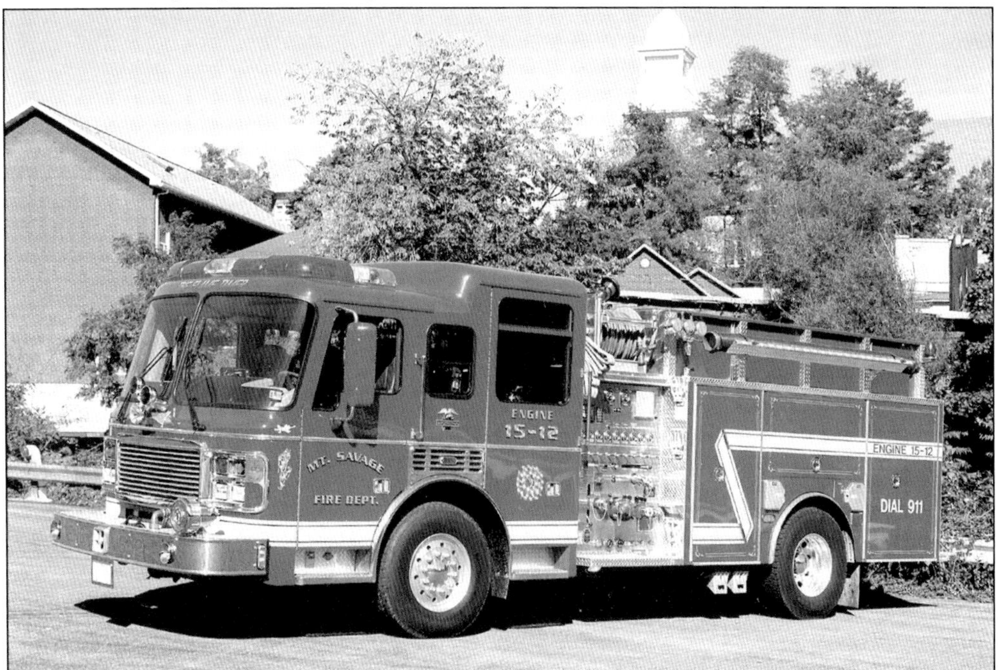

Mount Savage's most modern rig is this 1998 American LaFrance Eagle model/4 Guys pumper. The handsome apparatus can pump 1,500 GPM and carries 1,000 gallons of water. Mount Savage also has two International/4 Guys units, a rescue truck, and a tanker.

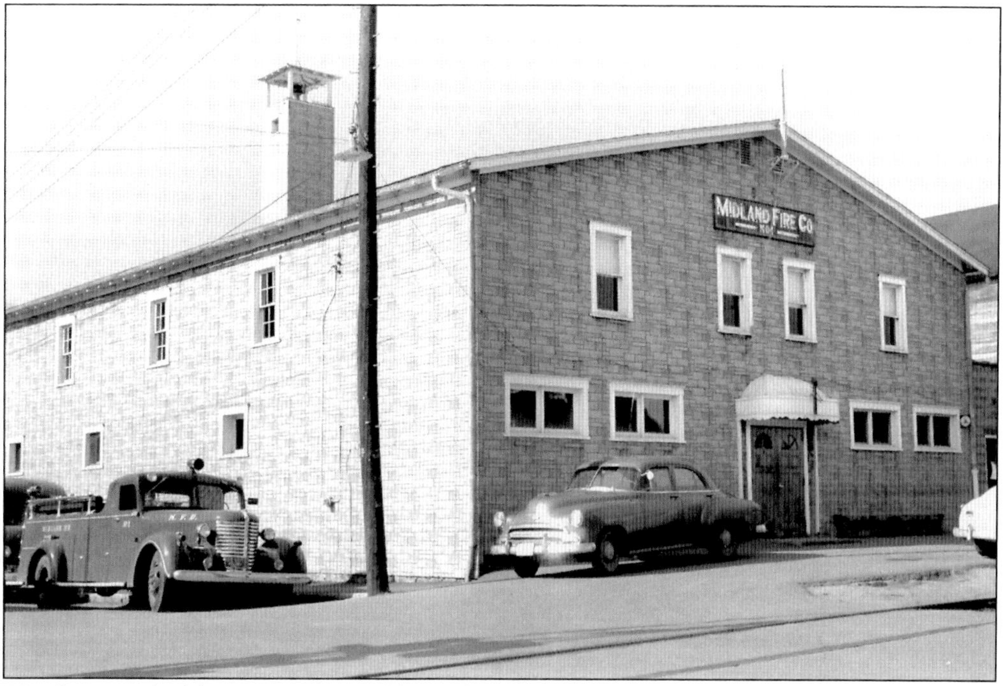

This business building was renovated by Midland VFC from 1931 to 1938 to create a social hall and meeting rooms. This is a photo from the 1950s with Midland's 1945 Buffalo pumper parked beside the structure. (Courtesy Ed Bosanko, MSFA.)

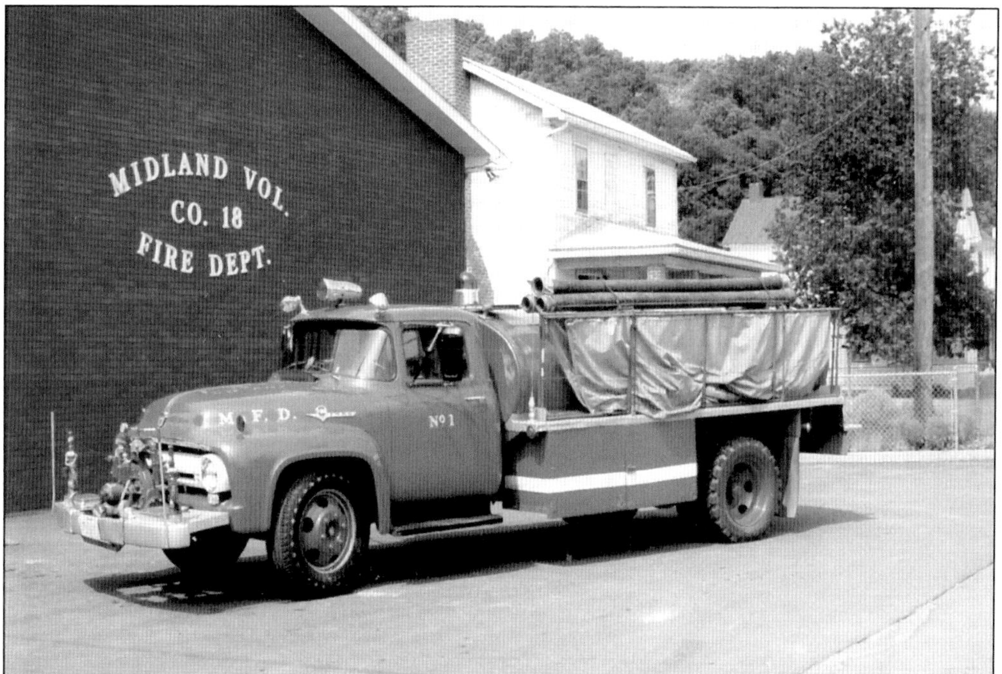

This 1956 Ford chassis was fitted with a 500-GPM pump and a 1,650-gallon water tank by the members of the fire company. This truck stayed in service until 1991. (Courtesy David Bowen.)

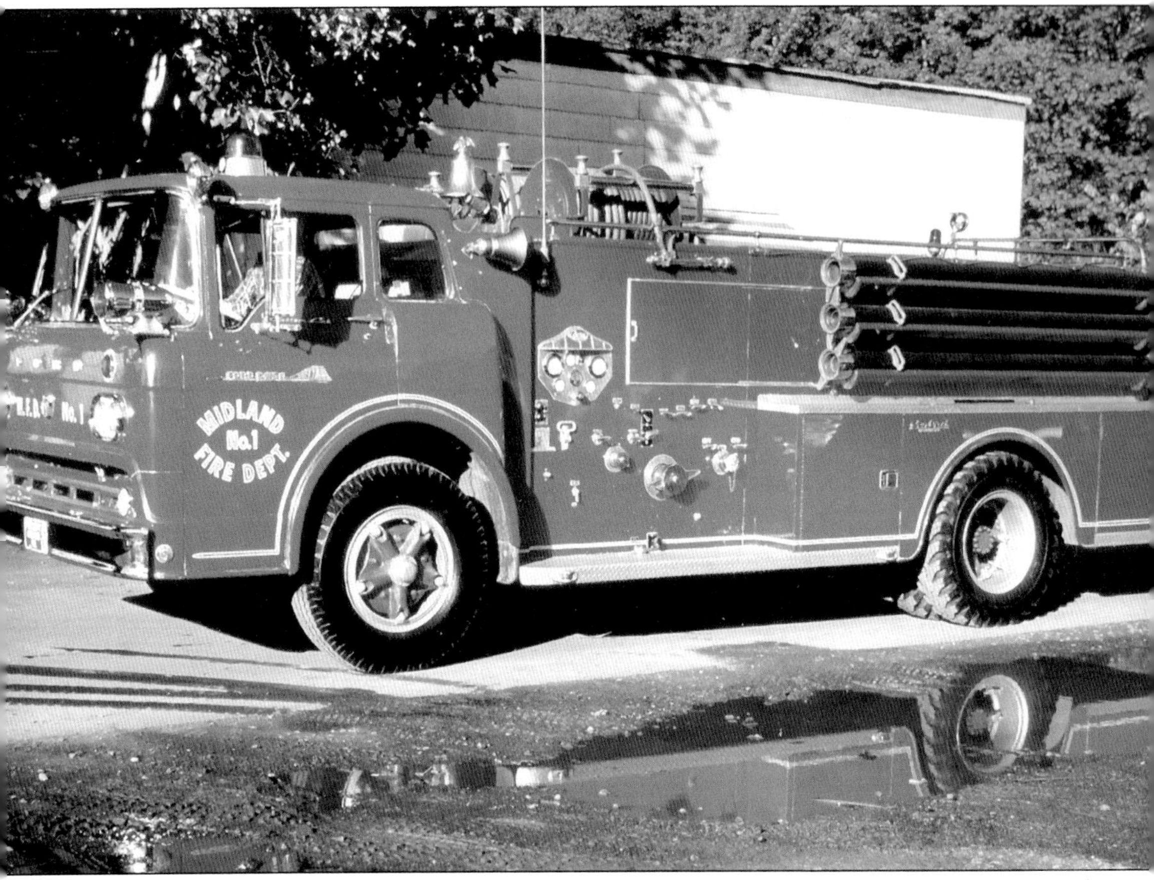

Midland took delivery in 1959 of this Ford C series/Central pumper with a 500-GPM pump and a 1,000-gallon water tank.

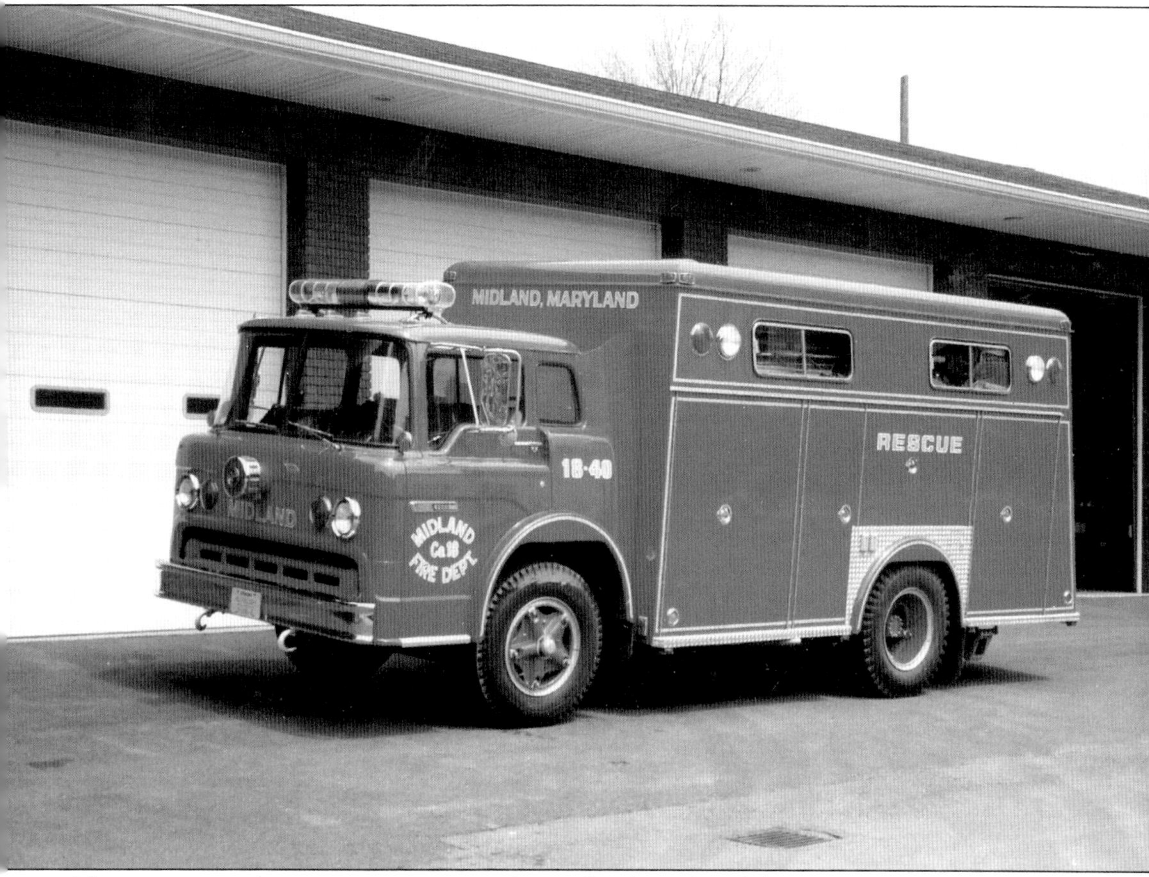

After using several step-vans as rescue trucks, Midland purchased this 1975 Ford C/Swab rescue truck, which carried an air cascade system, a five-kilowatt generator, and a jaws of life Hurst rescue tool. It is posed in front of the new engine house, which was built in 1981.

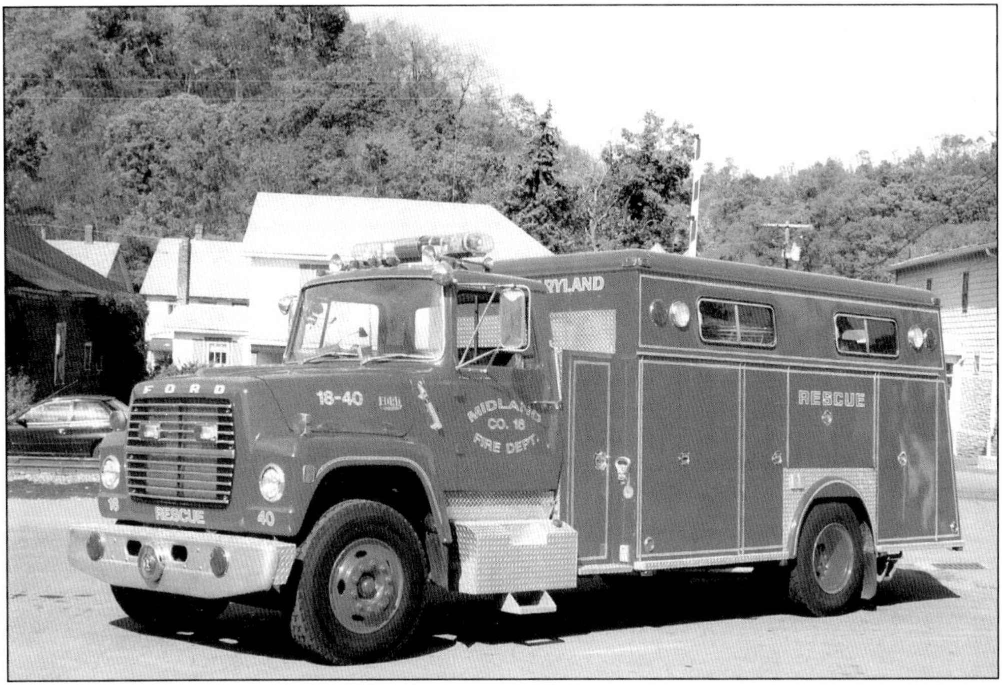

A 1971 Ford L series chassis from Bethesda (Maryland) Fire Department was used in 1992 to refurbish the 1975 Ford/Swab rescue truck. The box from the old chassis was transferred and other modifications were completed by 4 Guys.

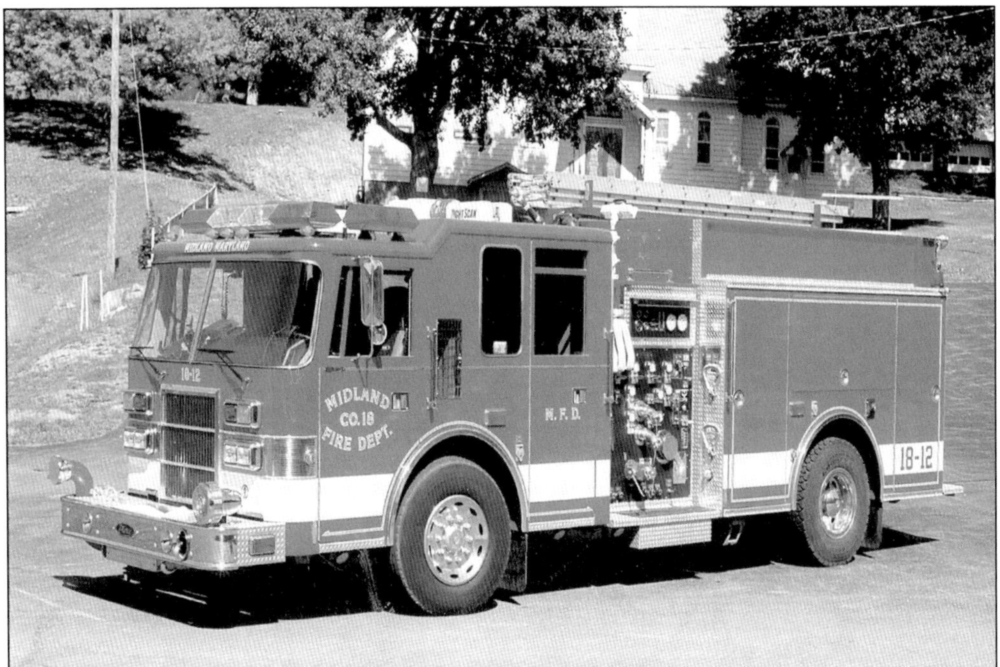

The 1967 Ford/Boardman pumper was replaced in 1998 with this Pierce Dash model with a 1,250-GPM pump and 1,000-gallon water tank. It is posed in front of Grace United Methodist Church, a Midland landmark for over 100 years.

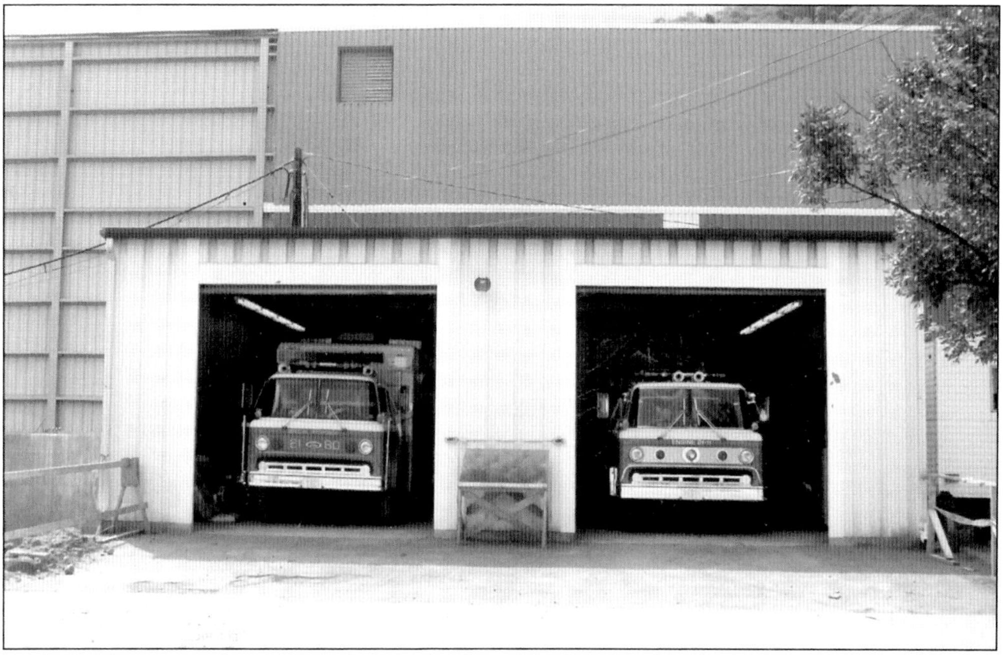

In the early 1980s, this fire station was built inside the paper mill property. Since most Luke volunteers work for the paper mill, this ensures a rapid response as the fire company is considered part of the paper mill's fire protection system. The apparatus are a 1990 Ford/4 Guys haz-mat squad (left) and a 1983 Ford/E-One 1,000-GPM pumper with a 750-gallon water tank.

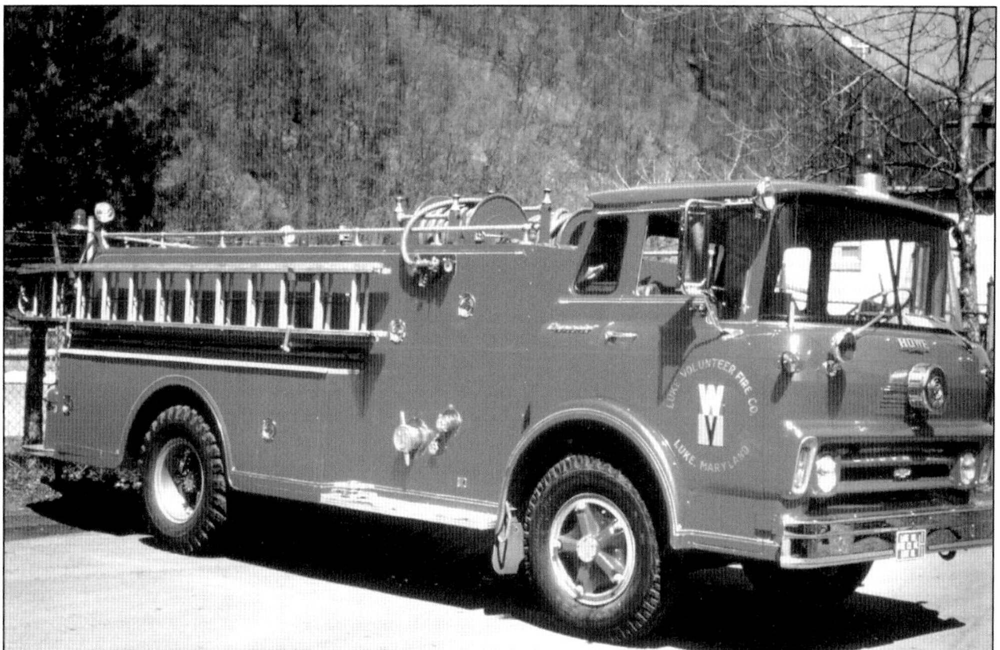

This 1965 Chevrolet/Howe 750-GPM pumper with a 500-gallon water tank was used by Luke VFC for 18 years until 1983. The "W/V" on the cab door is for Westvaco, Inc., then the owners of the paper mill.

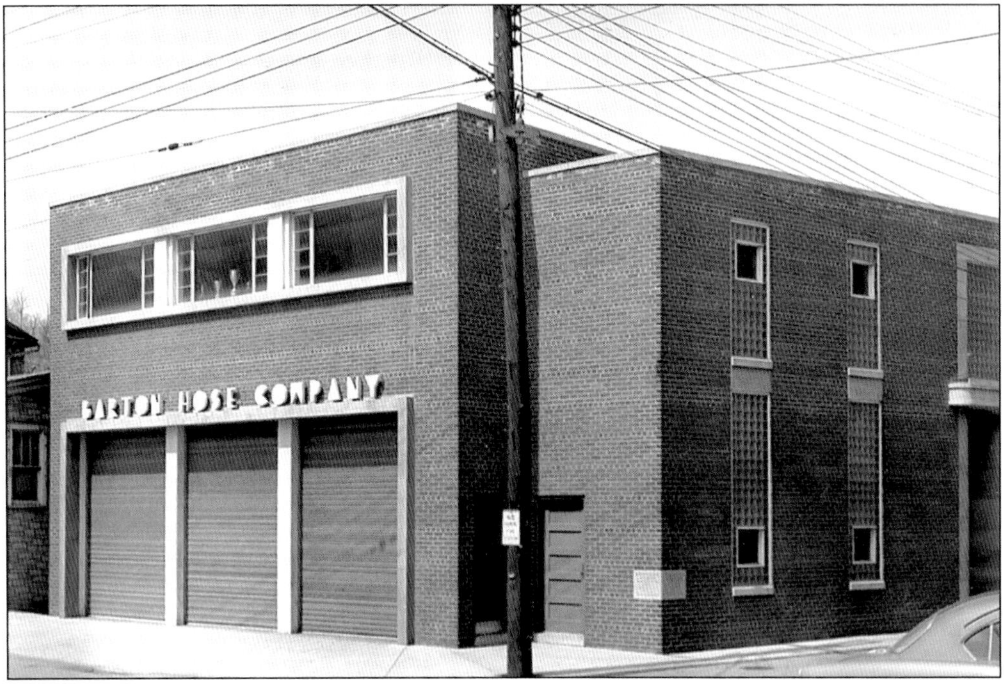

Here is Barton Hose Company's 1947 fire station as it appeared in the early 1950s. Note the stylized lettering on the front of the imposing brick structure that is a landmark in the town of Barton. (Courtesy Ed Bosanko, MSFA.)

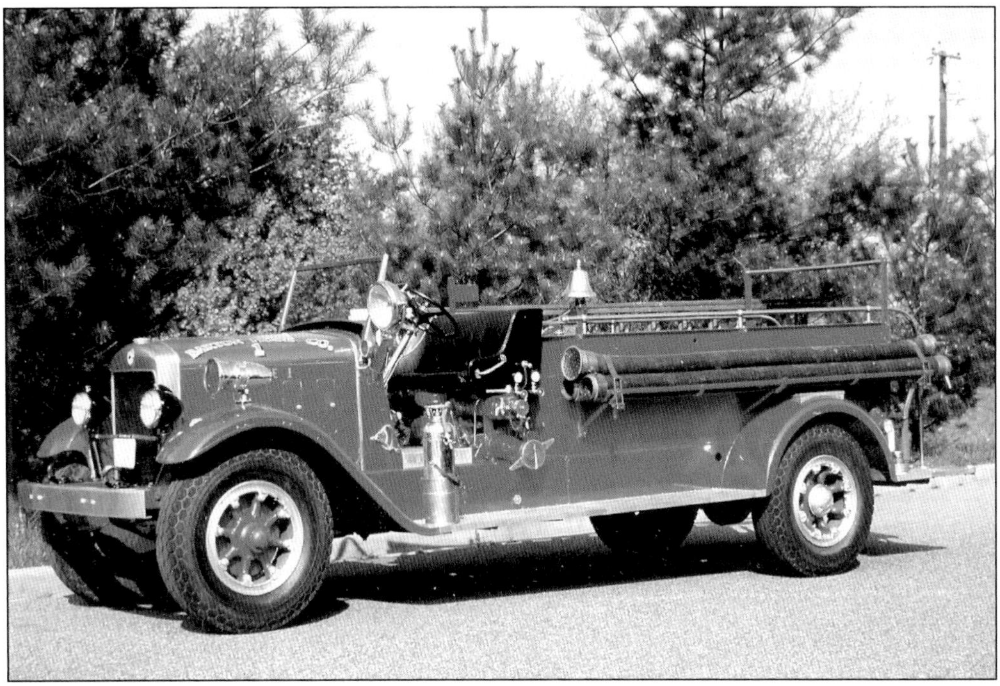

This 1932 Buffalo pumper was purchased in January 1933. It had a 500-GPM pump and carried 100 gallons of water. This truck replaced a 1927 Dodge chemical truck, which was Barton Hose's first motorized apparatus.

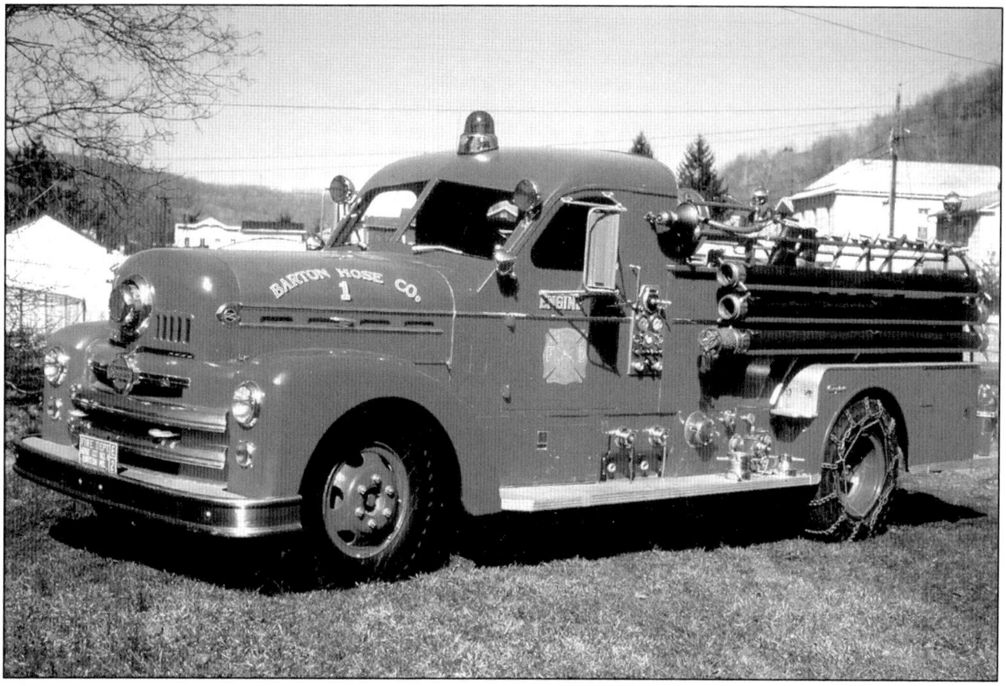

This 1952 Seagrave pumper could pump 750 GPM and probably carried at least 500 gallons of water. It would be painted yellow in the mid-1970s and then sold to Bowman's Addition VFC in the early 1980s.

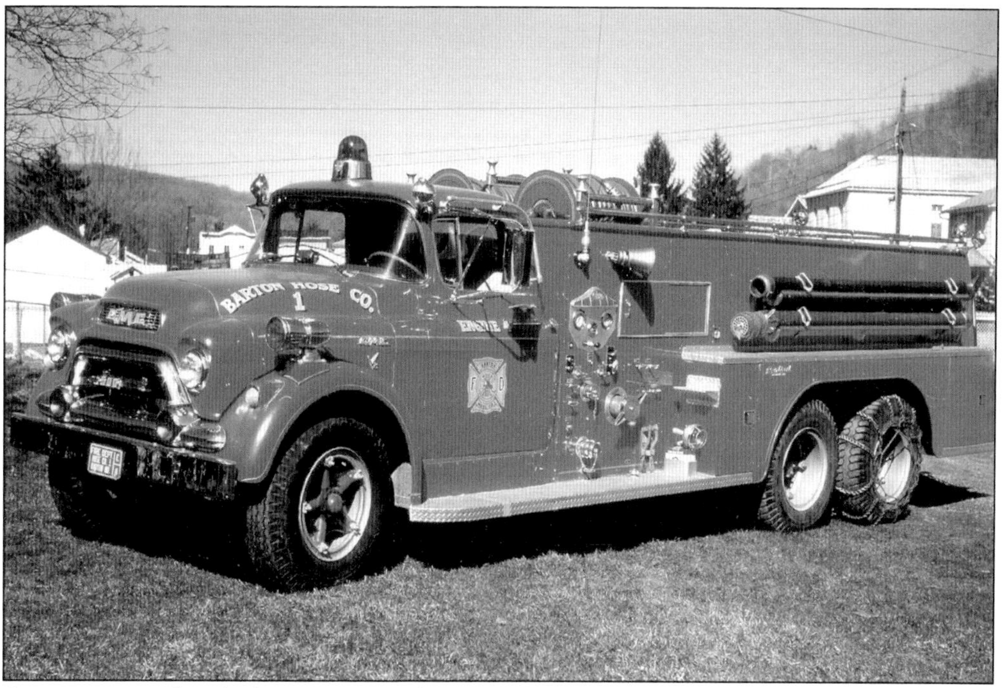

An apparatus ahead of its time was this 1958 GMC/Central pumper-tanker, which could pump 500 GPM and carry 1,600 gallons of water. The large apparatus was probably a maintenance nightmare.

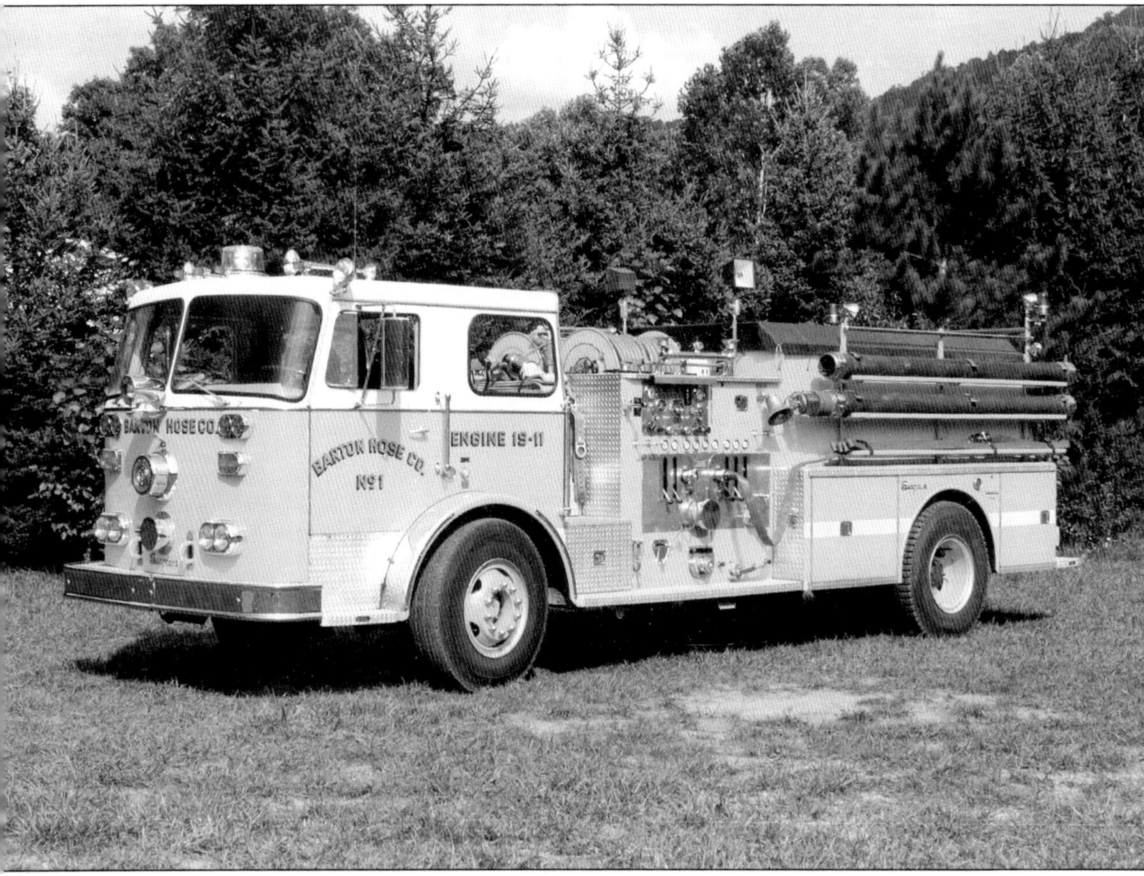

A Seagrave PB series pumper was delivered in 1975 in the new yellow paint scheme. It could pump 1,000 GPM and carry 1,000 gallons of water. It stayed in service for 25 years. (Courtesy Virginia McLaughlin.)

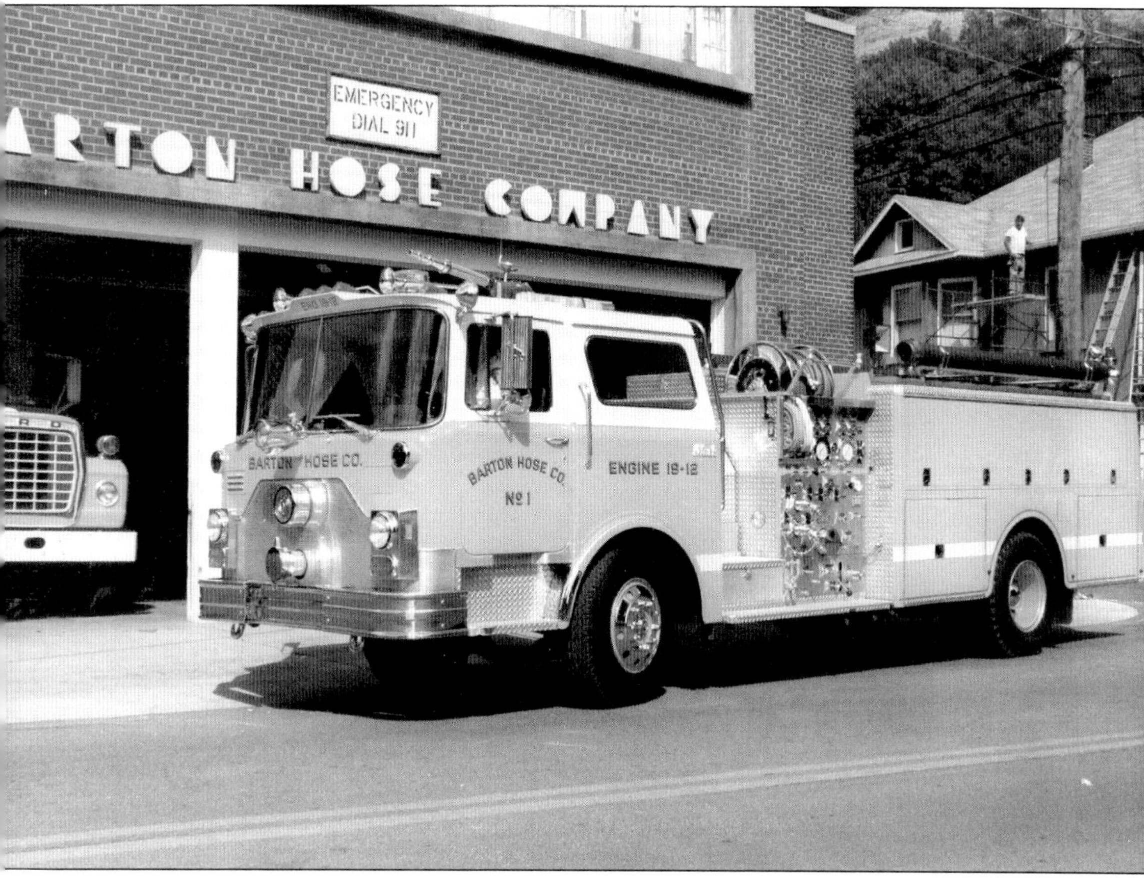

Shown in front of Barton Hose's station is a 1982 Mack CF series pumper that could pump 1,000 GPM and carry 1,000 gallons of water. (Courtesy Howard Meile.)

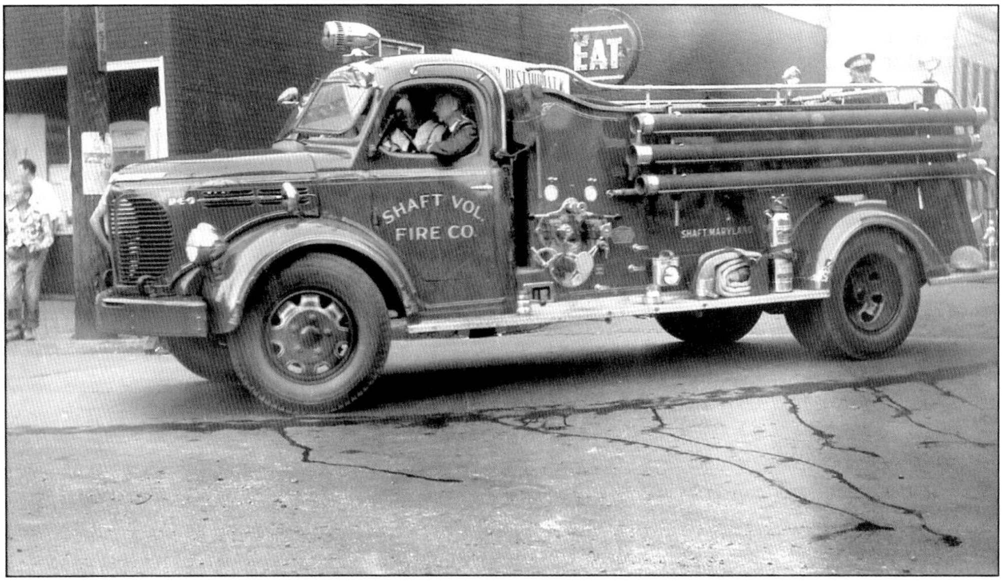

Shaft VFC's first new pumper, a 1947 REO/Oren, is shown in a parade. It could pump 500 GPM and carry 500 gallons of water. It was sold to Fountain (West Virginia) VFC in the early 1970s. (Courtesy William Snyder.)

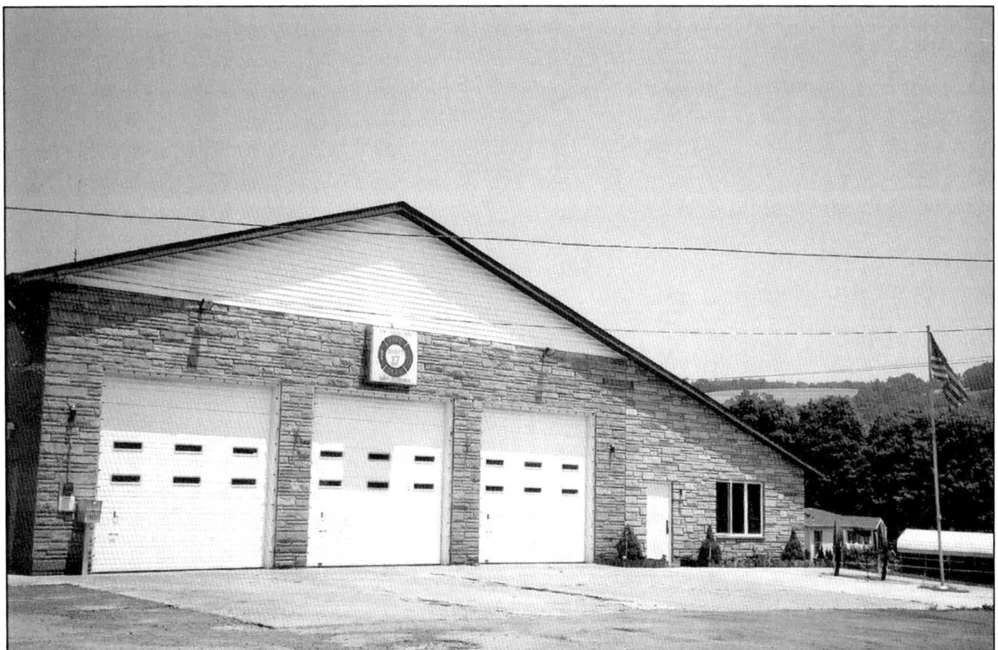

This large modern fire station was built in the mid-1960s and has been improved several times since. The original hose reel equipment of the fire company can be seen beside the flagpole.

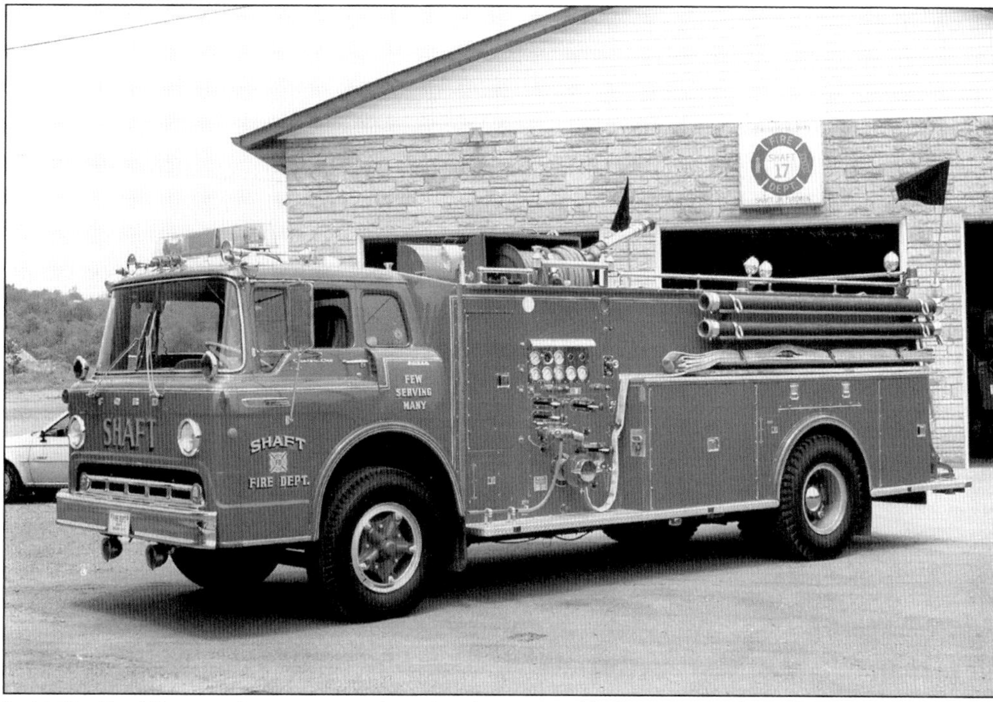

In 1968, a Ford C series chassis was purchased and completed by Boyer, an Indiana fire apparatus builder. It could pump 750 GPM and carry 1,100 gallons of water. (Courtesy Virginia McLaughlin.)

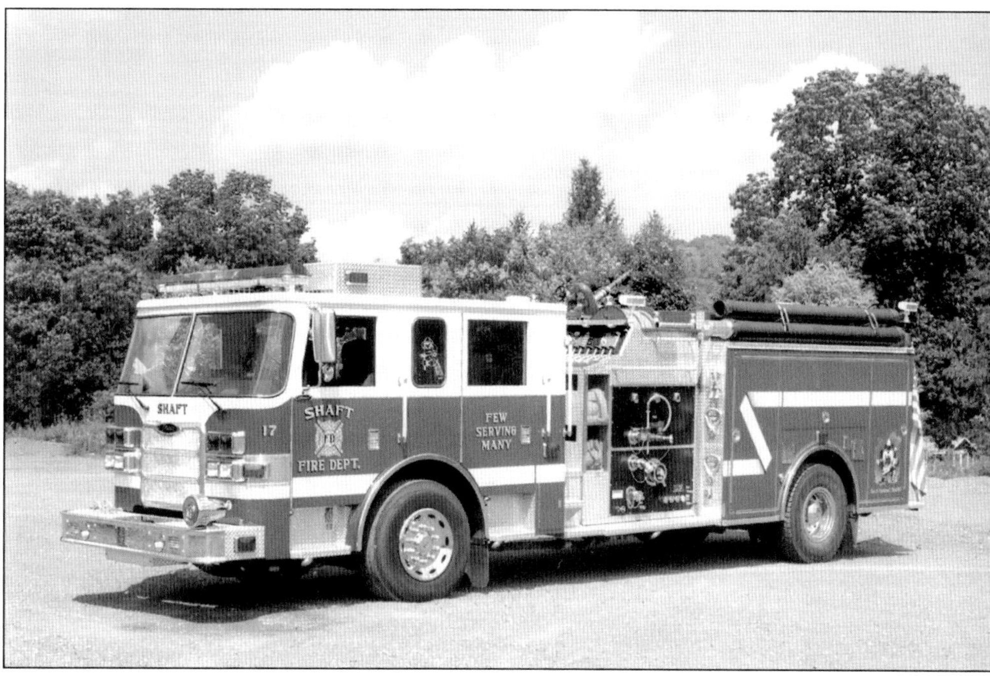

This Pierce Arrow XT model pumper was delivered in 2004. It can pump 1,250 GPM and carry 750 gallons of water. This truck was honored with a photo on the 2005 Pierce official calendar.

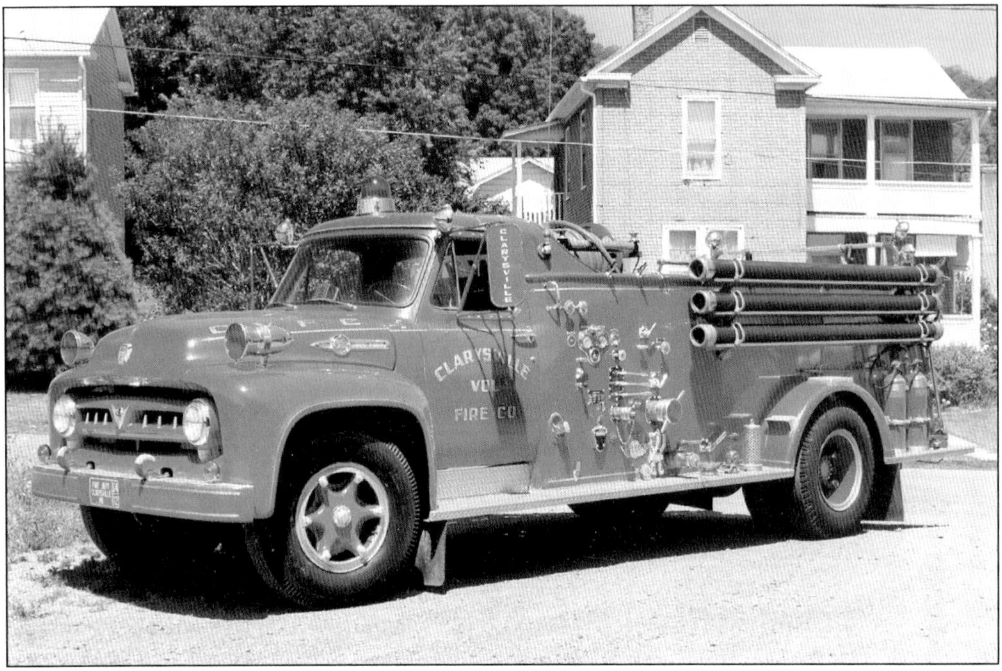

Clarysville VFC's first new fire truck was this 1953 Ford/American pumper, which had a 500-GPM pump and a 500-gallon water tank. This truck replaced a 1943 International pumper that was purchased from army surplus in 1946. (Courtesy Virginia McLaughlin.)

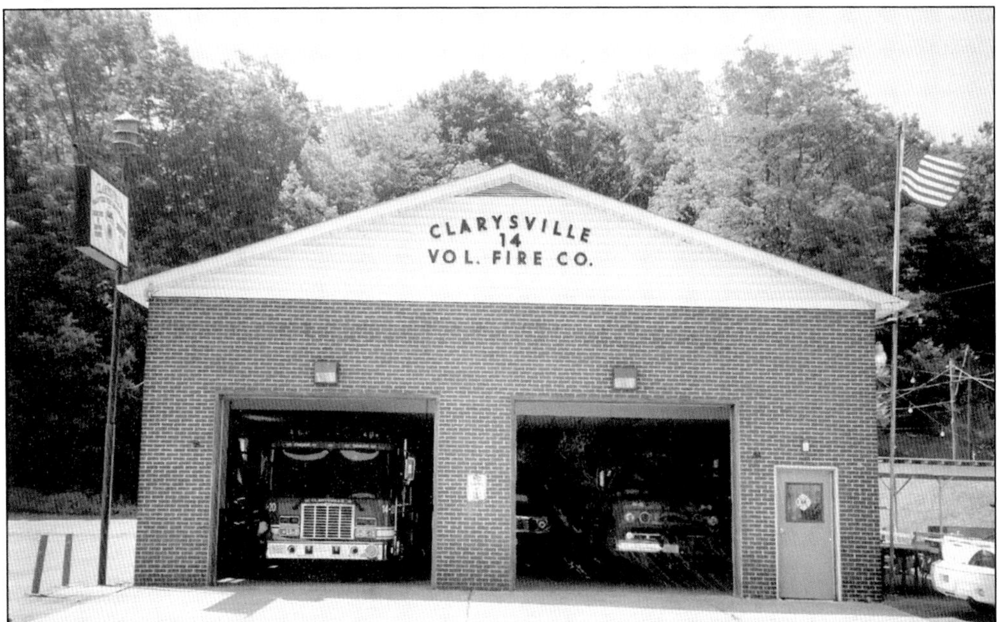

This is Clarysville VFC's present station with current apparatus visible inside. This building had an addition in 1973 for a social hall and kitchen, but the engine bays have been unchanged since 1957. Apparatus visible from left to right are a 1997 HME/4 Guys pumper tanker, a 1977 Dodge brush truck, and a 1988 Ford C/4 Guys pumper.

Four

Rural and Suburban Firefighting

Prior to the 1930s, fire protection was almost non-existent in many areas of Allegany County, outside Cumberland, and outside the towns along Georges Creek. As highways were improved between Frostburg and Cumberland and other points in the county, these departments would respond but were often hampered by low water supply.

LaVale grew rapidly as the National Road was improved into U.S. Route 40. In 1933, local citizens formed LaVale VFC and converted a Lincoln touring car into a fire apparatus.

Cresaptown, a town about six miles from Cumberland on the McMullen Highway to Keyser, West Virginia, organized a volunteer fire company in 1933.

The village of McCoole is located on the McMullen Highway (U.S. Route 220) over 20 miles south of Cumberland. Its fire company was formed in 1938 and obtained a used 1918 American LaFrance pumper from the Baltimore area as the first apparatus.

Corriganville is a small village about four miles northwest of Cumberland. The citizens formed a fire company in 1939.

Baltimore Pike, an area about five miles east of Cumberland on U.S. Route 40, formed a fire company in May 1942.

Bowman's Addition, an area just north of Cumberland, organized as an auxiliary Civil Defense fire brigade in April 1942.

Flintstone, another village on U.S. Route 40, about 12 miles east of Cumberland, formed their fire company in August 1943.

Bedford Road, a residential area north of Cumberland on U.S. Route 220, formed an auxiliary fire company with assistance from Civil Defense and Cumberland Fire Department in November 1943.

District 16 VFC, named for the county election district, which covers the area southeast of Cumberland, was organized in July 1944.

The Ellerslie VFC was organized in November 1946. By April 1947, the new company had obtained an alerting siren and a used fire truck from Cumberland and had begun construction of their first station.

In June 1947, the Bowling Green VFC was formed, and their first fire station was built about three miles south of Cumberland on U.S. Route 220.

Oldtown VFC was formed in 1958 and used a wooden shack for a station until a bigger station was built on Maryland Route 51. The volunteers obtained two used fire trucks to serve as their first apparatus.

The rural area of Orleans in the far eastern part of Allegany County formed a fire company in 1980.

The newest fire company in Allegany County was formed in the village of Rawlings about midway between Cresaptown and McCoole on U.S. Route 220.

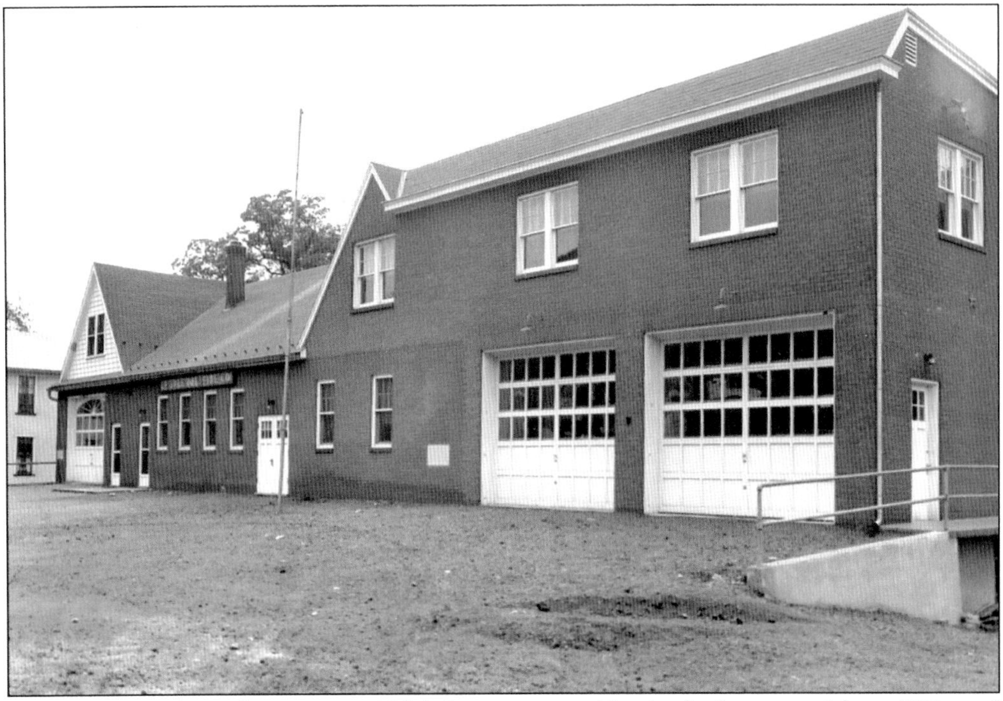

Shown here in the early 1950s is LaVale's former station No. 1, which was used from 1938 until 1997. It is now used for fund-raising activities that are held in the social hall. (Courtesy Ed Bosanko, MSFA.)

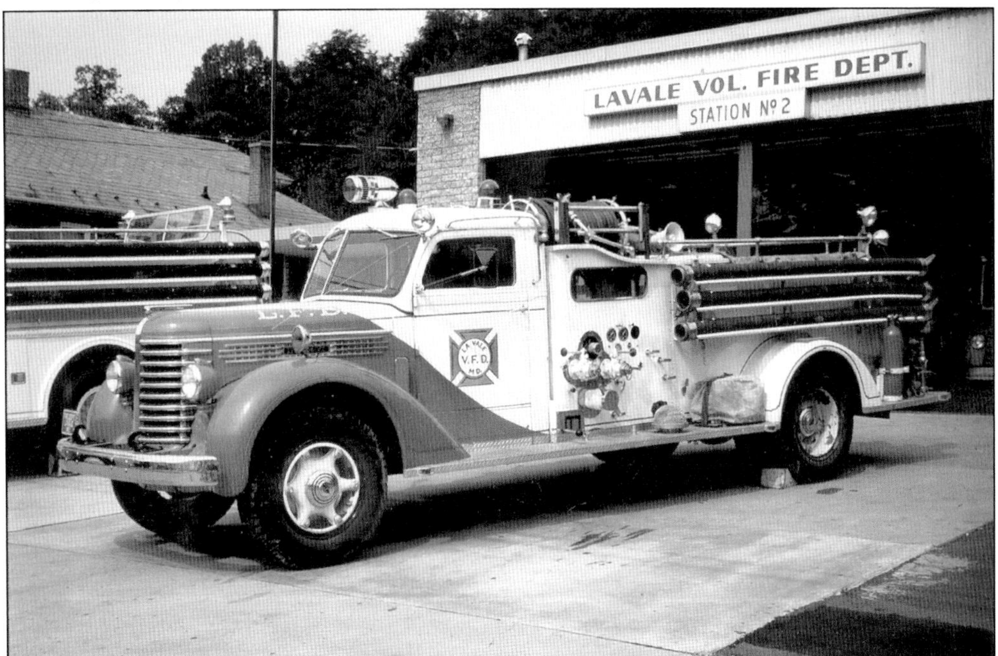

The LaVale VFD had two of these Diamond T/Oren pumpers for many years. This 1949 model had a 500-GPM pump and carried 500 gallons of water. It was donated to the Orleans VFC in 1980. LaVale Station No. 2 was built in 1965.

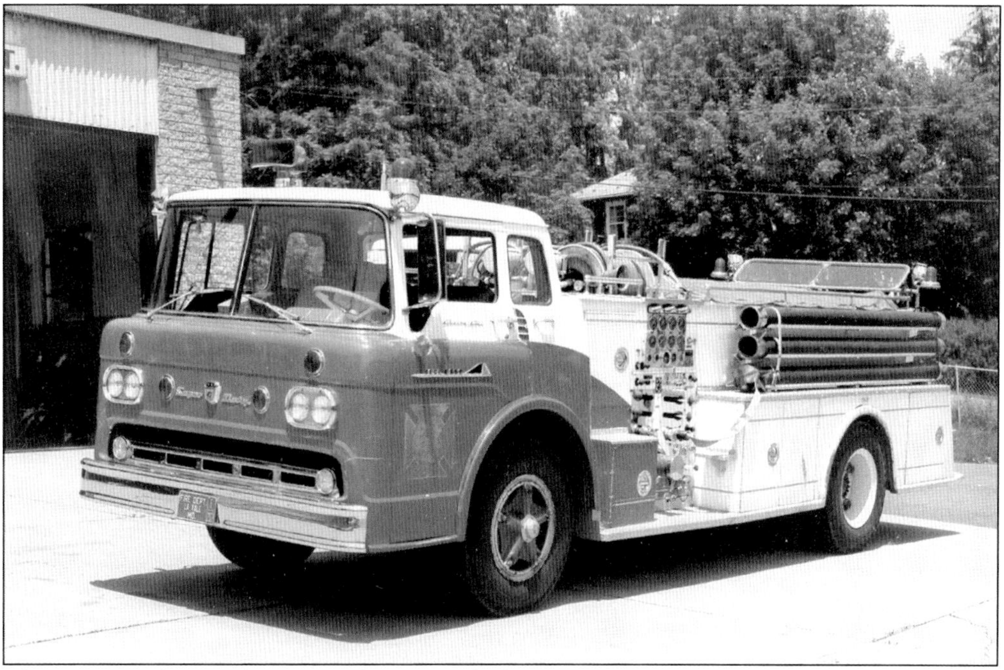

This 1959 Ford C-850/Peter Pirsch pumper had a 750-GPM pump and carried 500 gallons of water. It served until the early 1990s. (Courtesy Virginia McLaughlin.)

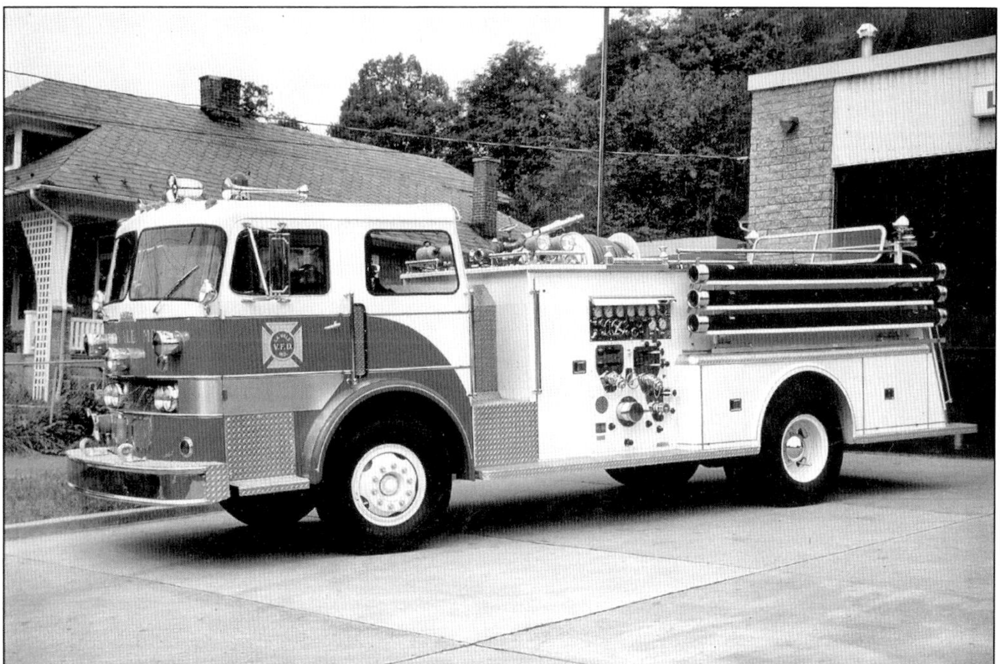

The first diesel fire apparatus in the LaVale fleet was this 1968 Bruco custom pumper that has a 750-GPM pump and carries 600 gallons of water. It is still owned by the LaVale VFD.

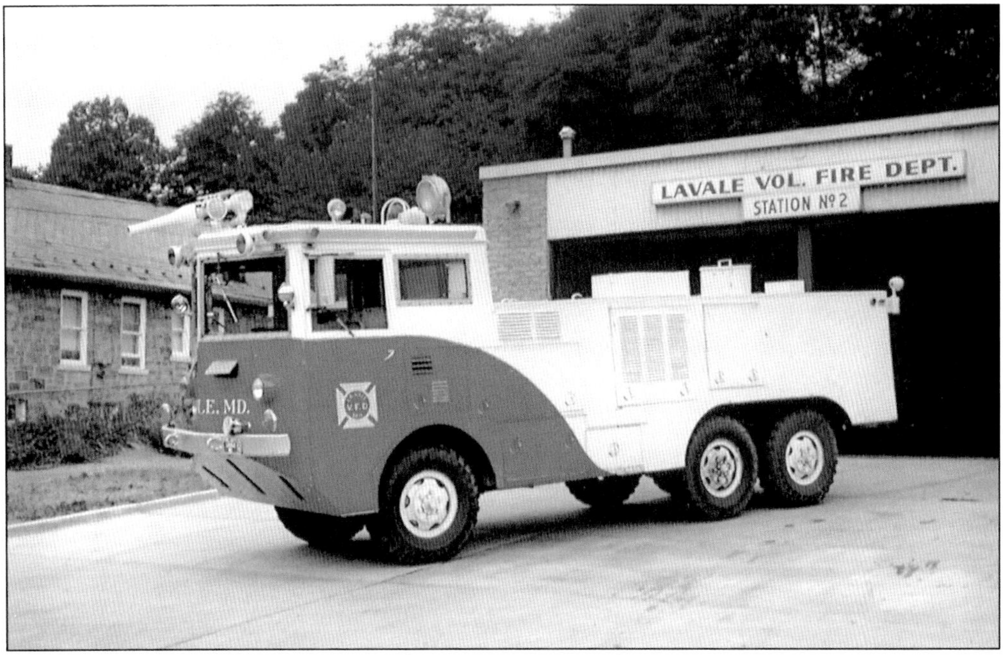

Although LaVale is a few miles from the nearest airport, this 1952 American LaFrance airport crash truck was acquired to deal with any highway disasters on the new freeway passing through the area. The big and unusual apparatus only served a few years during the 1960s and 1970s, but it looked good in parades. (Courtesy Jeff Mogush.)

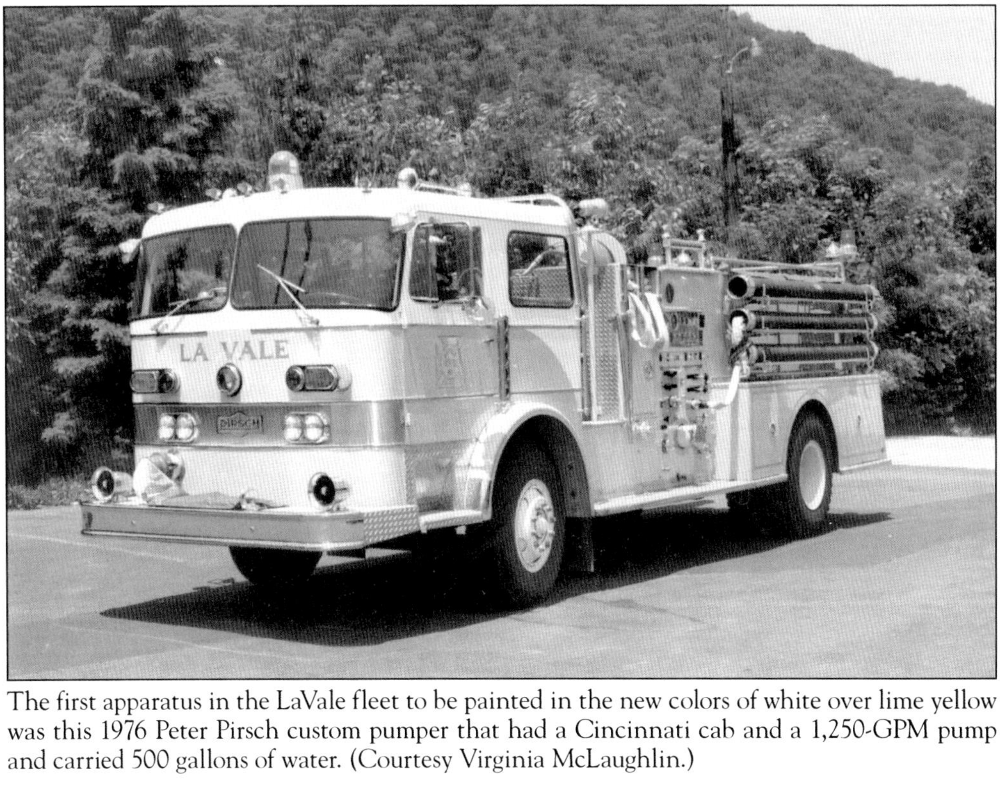

The first apparatus in the LaVale fleet to be painted in the new colors of white over lime yellow was this 1976 Peter Pirsch custom pumper that had a Cincinnati cab and a 1,250-GPM pump and carried 500 gallons of water. (Courtesy Virginia McLaughlin.)

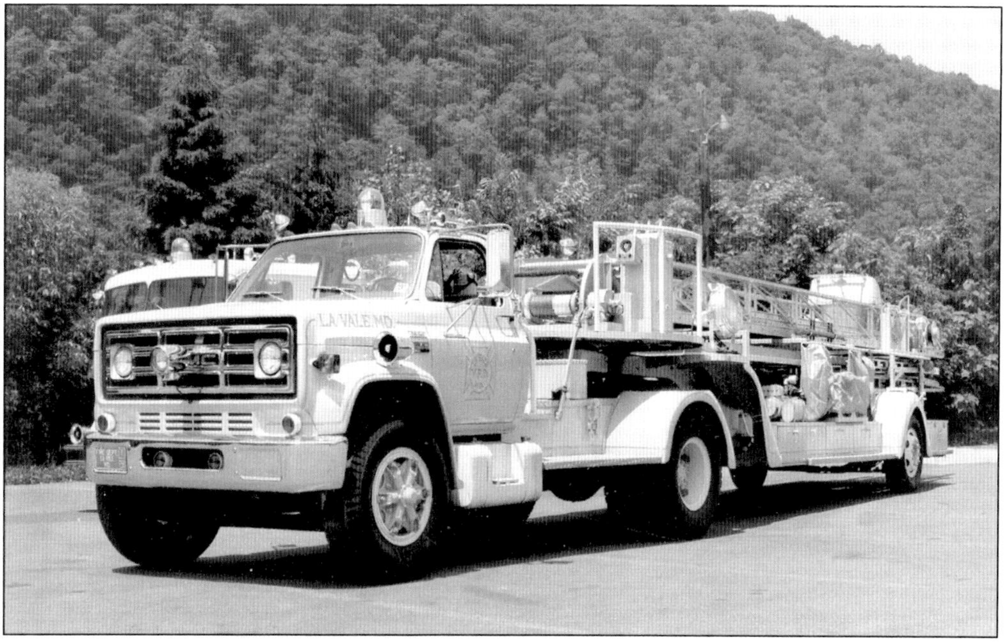

The first aerial ladder apparatus used by LaVale was this 1945 Seagrave, which was originally purchased by Cumberland from Easton, Maryland. LaVale purchased the rig in 1977 from Cumberland, Maryland. In 1978, J and J Truck Bodies in Somerset, Pennsylvania, replaced the original Seagrave tractor with a 1978 GMC and repainted the entire truck in the white-over-lime-yellow scheme. (Courtesy Virginia McLaughlin.)

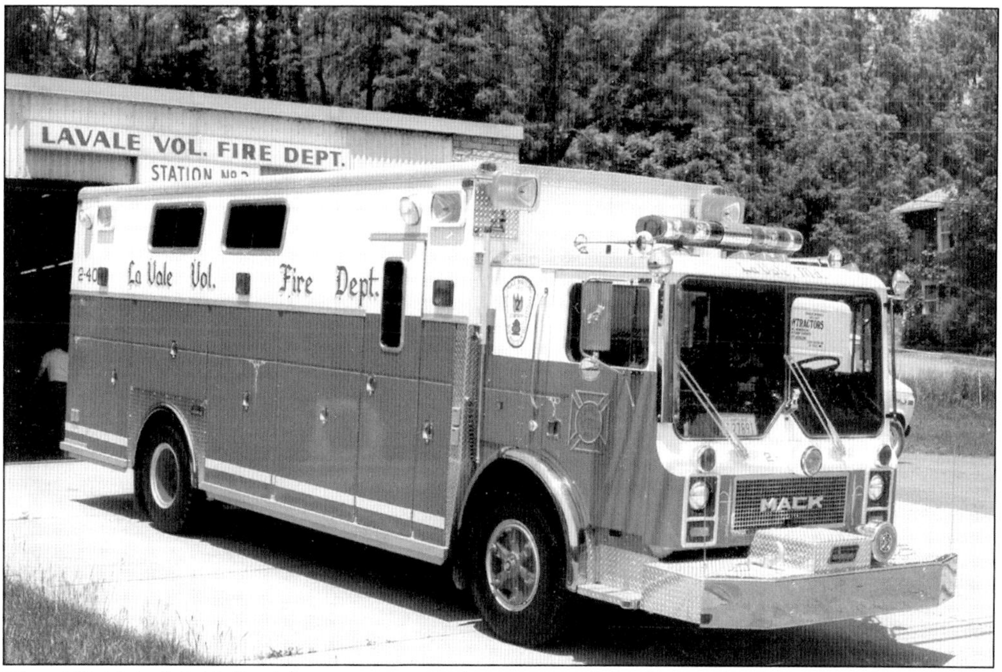

This 1984 Mack MC/Saulsbury rescue truck was one of the largest built by the manufacturer up to that time. This truck brought back white-over-red colors to the LaVale VFD. (Courtesy Virginia McLaughlin.)

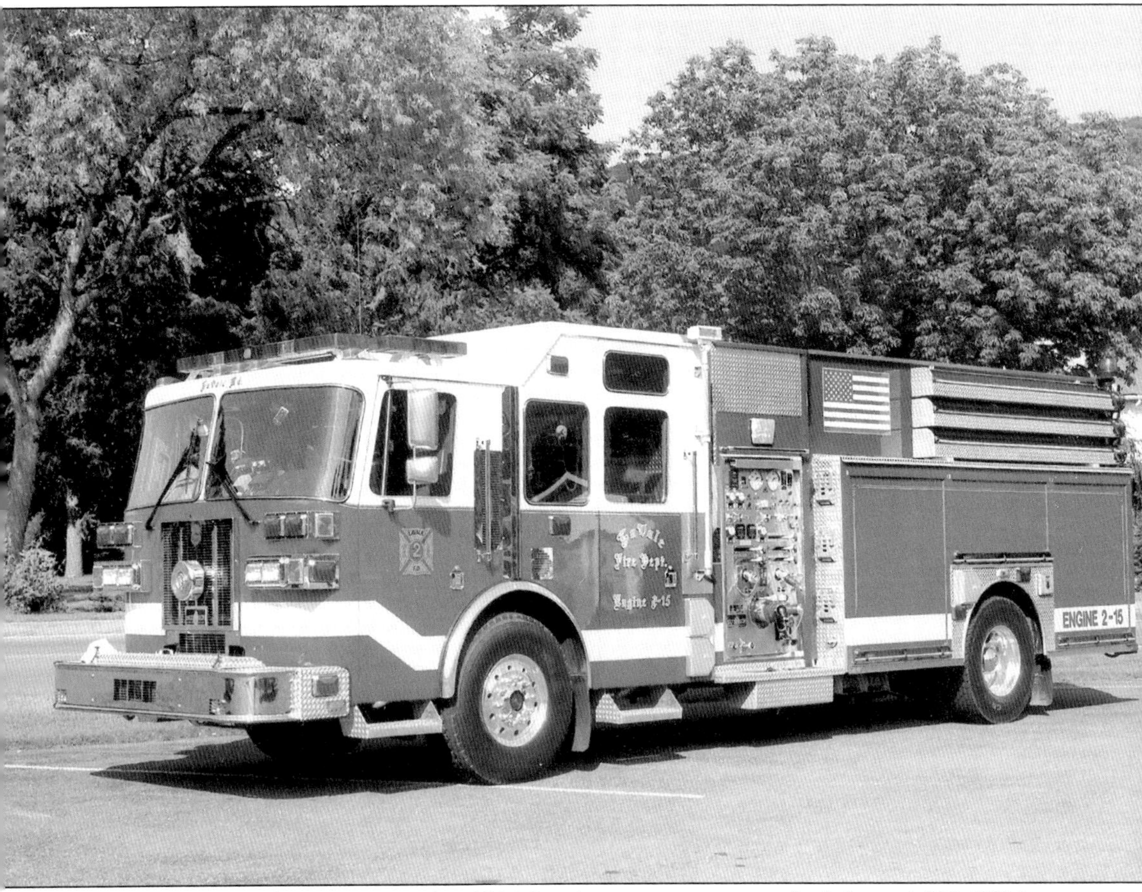

The newest LaVale apparatus is this 2004 Sutphen rescue engine, which has a 1,500-GPM pump with a 500-gallon water tank.

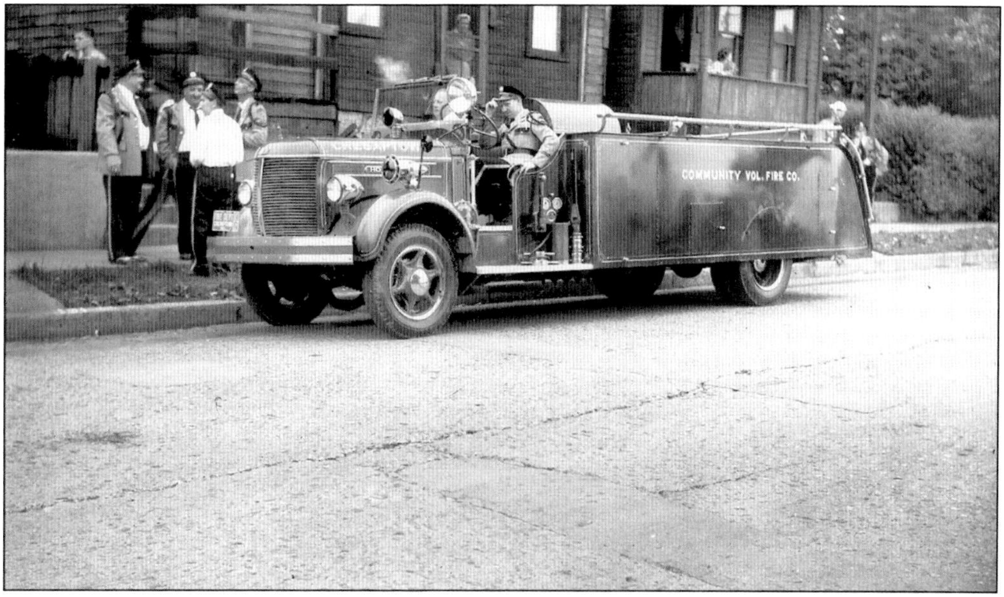

Cresaptown's first new pumper was this 1940 Howe Defender custom unit, which was in service into the mid-1960s. The unit is seen at a parade in 1951 with the firemen in formal dress uniform. (Courtesy William Snyder.)

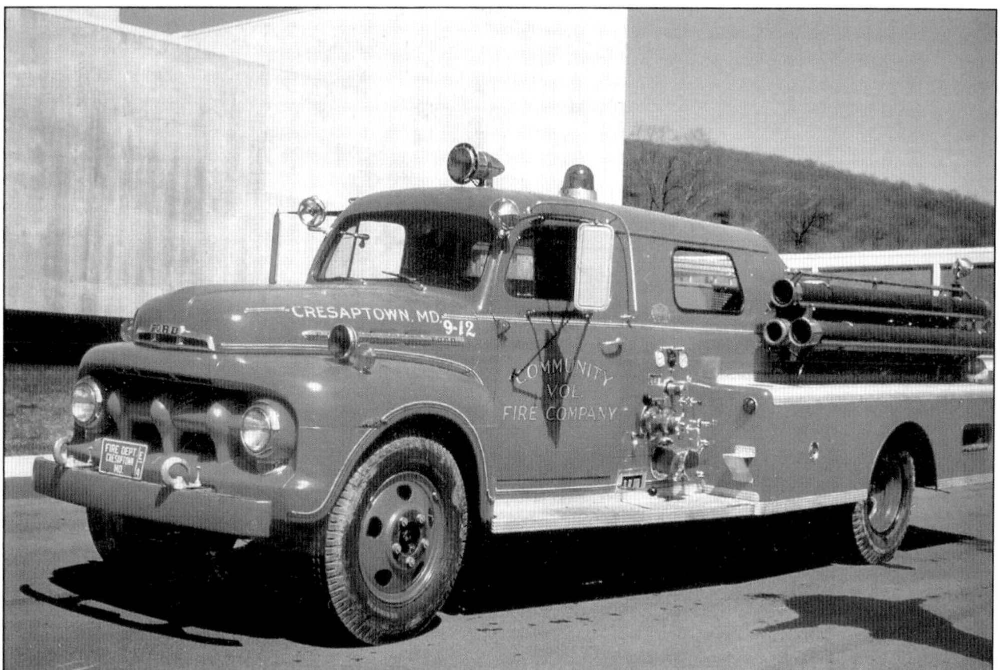

The next new pumper was this 1951 Ford/Oren with a 750-GPM pump. It stayed in service until at least 1976. Note the unusual placement of the booster reel in the rear compartment and the covered canopy over the crew seat, which was reached via a walkway from the tailboard.

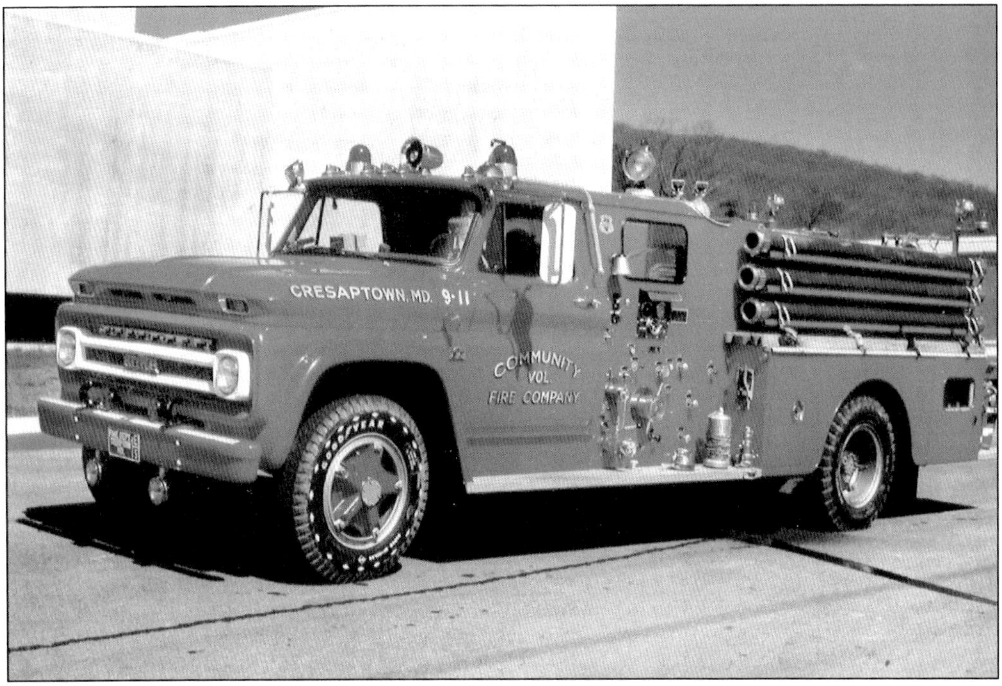

This 1966 Chevrolet/Howe had a 750-GPM pump and carried at least 500 gallons of water. Like the 1951 Ford/Oren, this unit also had booster reels in the rear compartments on both sides and the canopy cab.

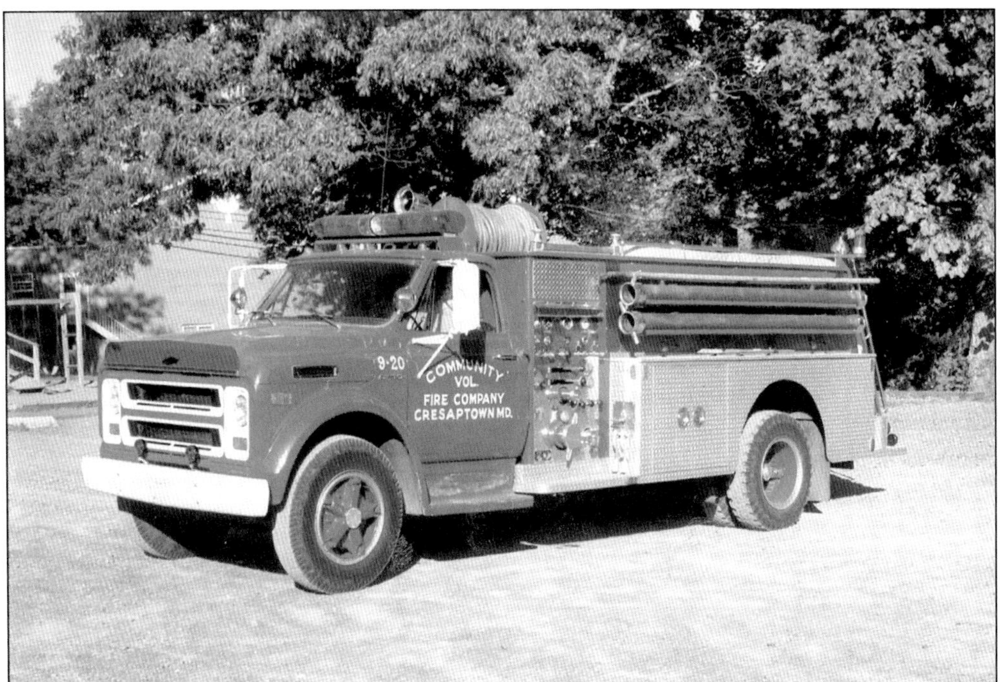

In 1972, Cresaptown replaced an older tanker with this Chevolet/Stinebaugh tanker, which had a 500-GPM pump and carried 1,500 gallons of water. (Courtesy John Floyd.)

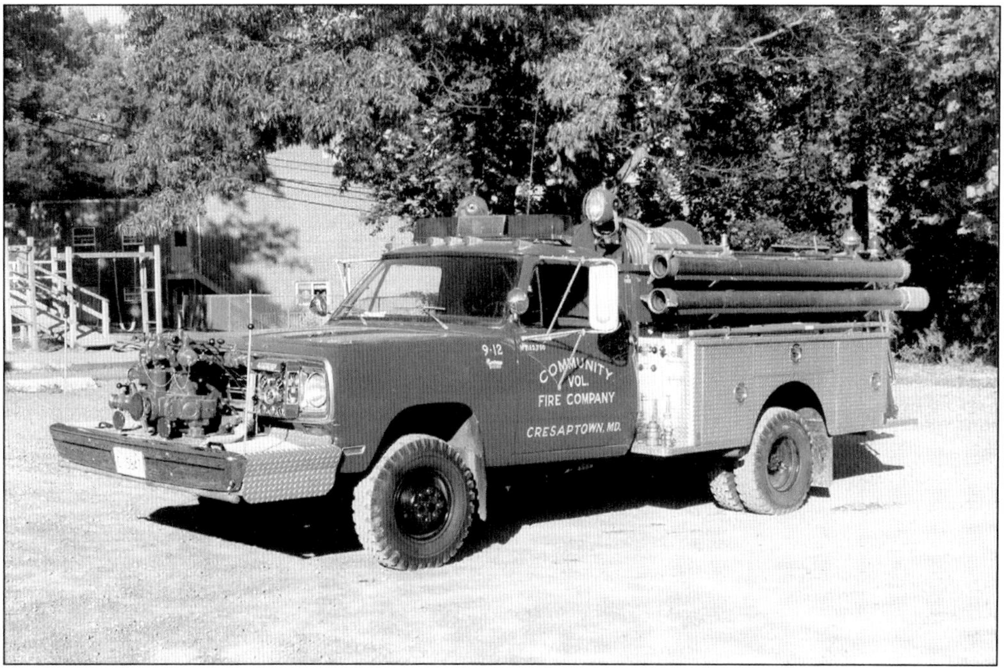

Cresaptown's second Stinebaugh delivery was this 1976 Dodge 4x4 unit that had a 750-GPM pump and carried 330 gallons of water. Stinebaugh apparatus were built in nearby Great Cacapon, West Virginia, and were marketed in four states. (Courtesy John Floyd.)

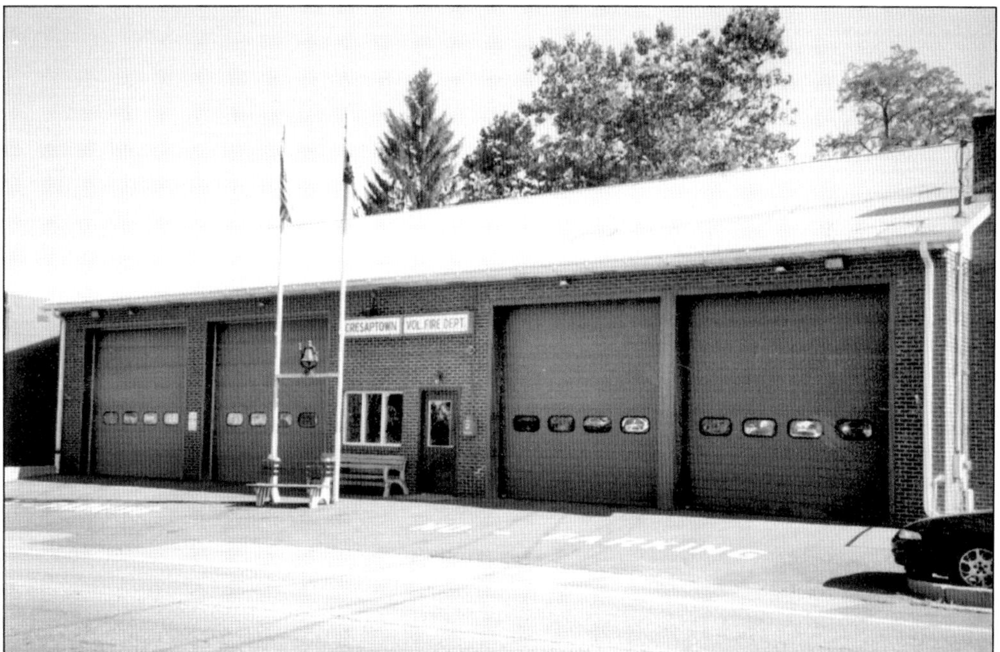

Here is Cresaptown's fire station as it currently exists. The original building was built in 1948 and was enlarged in 1986 and in 1998. The bell in front of the station is from an old C&P Railroad steam engine. Cresaptown's current apparatus fleet consists of three Pierce Lance models.

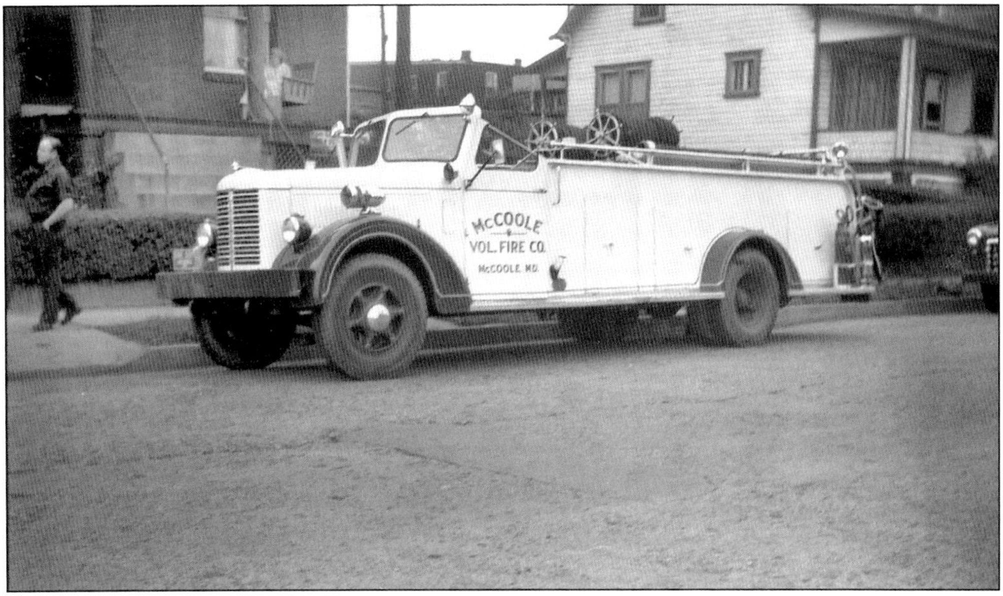

The 1949 Oren pumper was built on a rugged Available truck chassis. This apparatus stayed in service until 1979 and had a 500-GPM pump and carried 500 gallons of water. It was donated in 1986 to the new Rawlings VFC. (Courtesy William Snyder.)

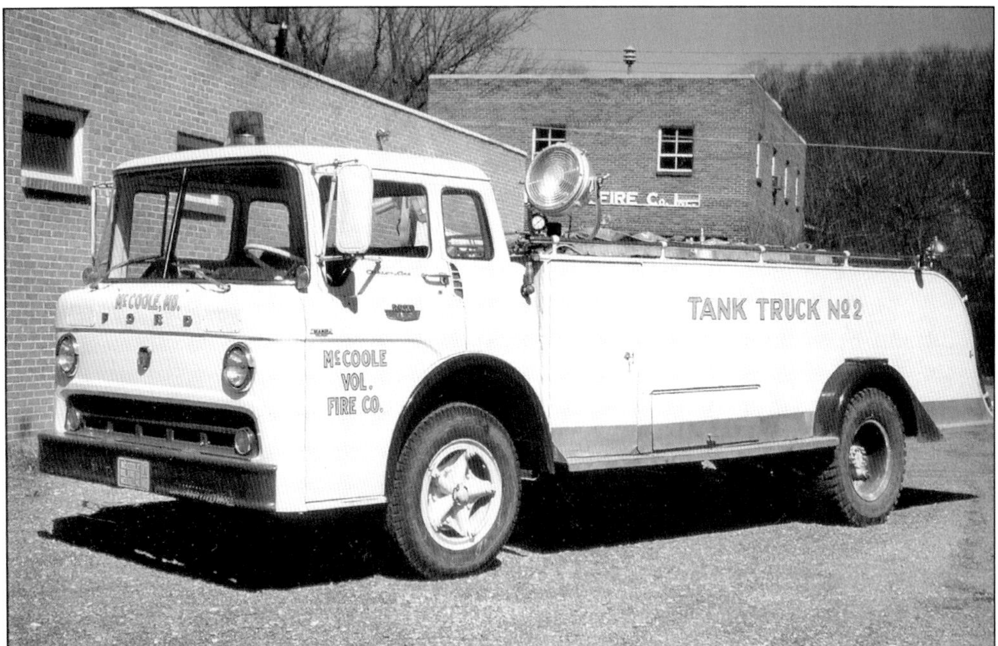

A 1964 Ford C series truck chassis was combined with a 1941 squad truck body (from Westernport), outfitted with a 1,000-gallon water tank and a small pump. This hybrid unit stayed in service until 1980. The old fire house is shown behind the truck.

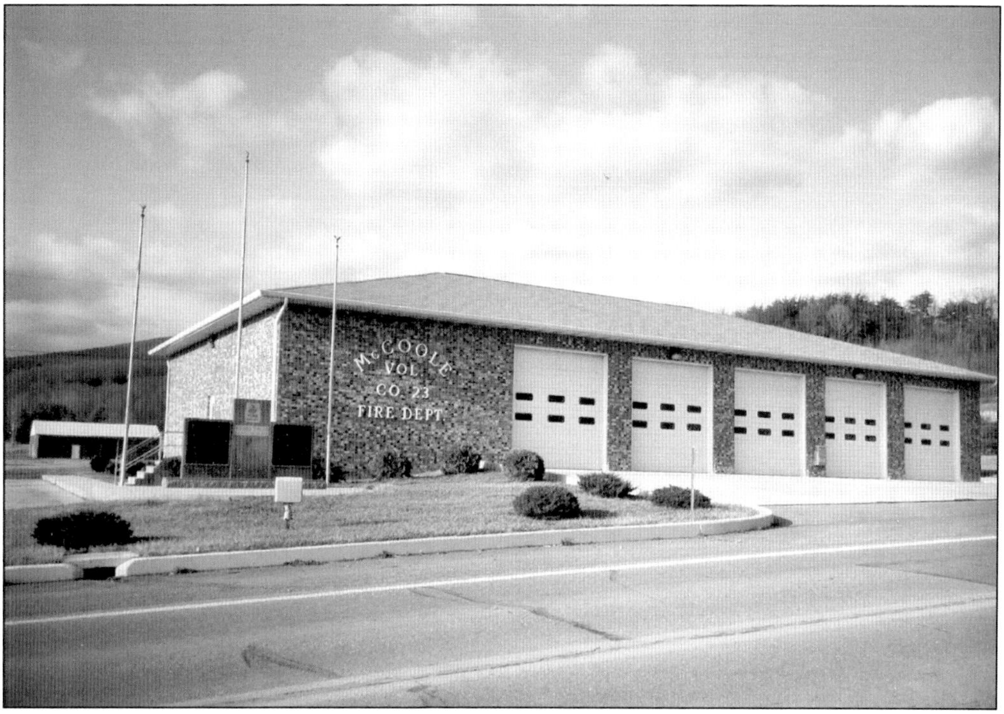

In December 1984, McCoole VFC occupied this modern five-bay fire station about one-half mile from the old station.

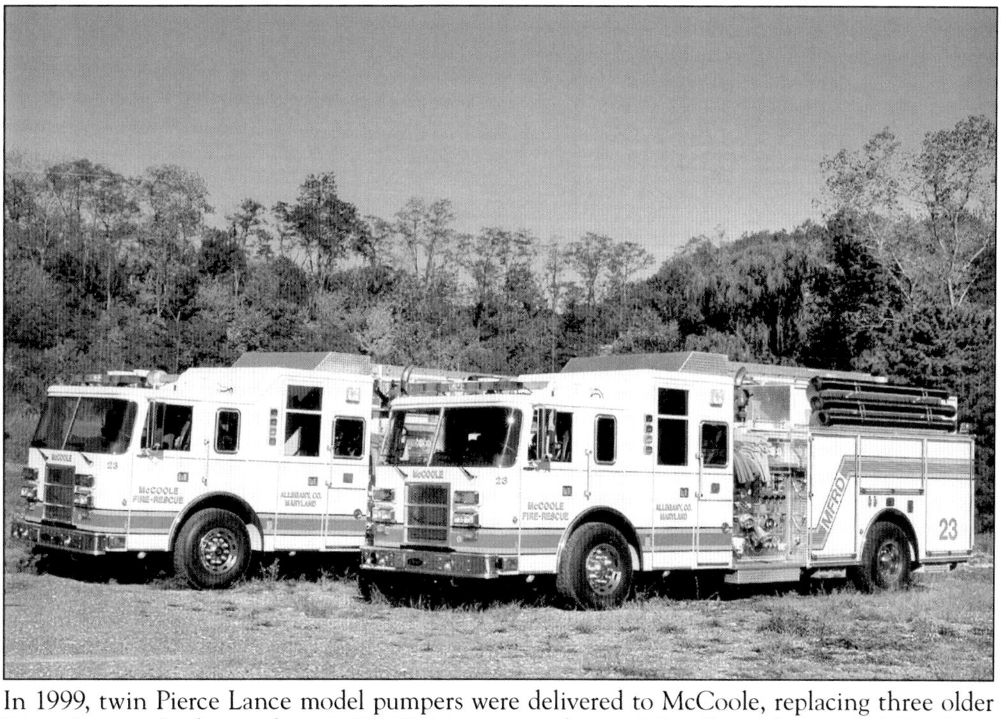

In 1999, twin Pierce Lance model pumpers were delivered to McCoole, replacing three older Pierce Lances. Both units have 1,500-GPM pumps and carry 750 gallons of water and 55 gallons of foam. One of the pumpers carries rescue equipment for auto accidents.

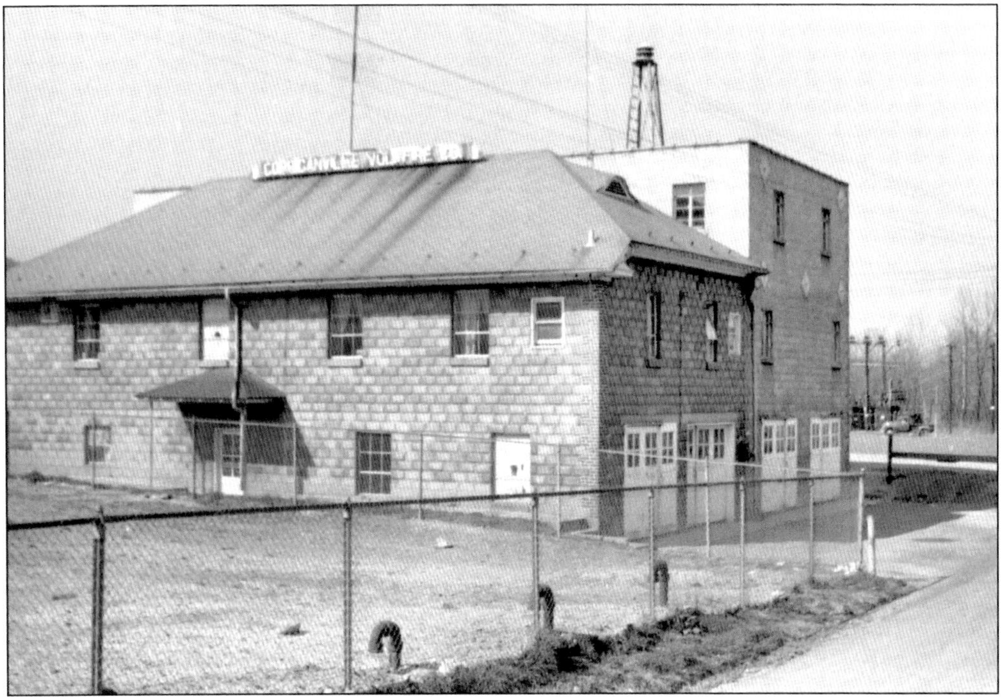

Corriganville's fire hall was built just before World War II. This photo is from the early 1950s, after an addition had been built. This station was used until the early 1980s, when a former elementary school was converted into apparatus bays, meeting rooms, and offices for the firefighters. (Courtesy Ed Bosanko, MSFA.)

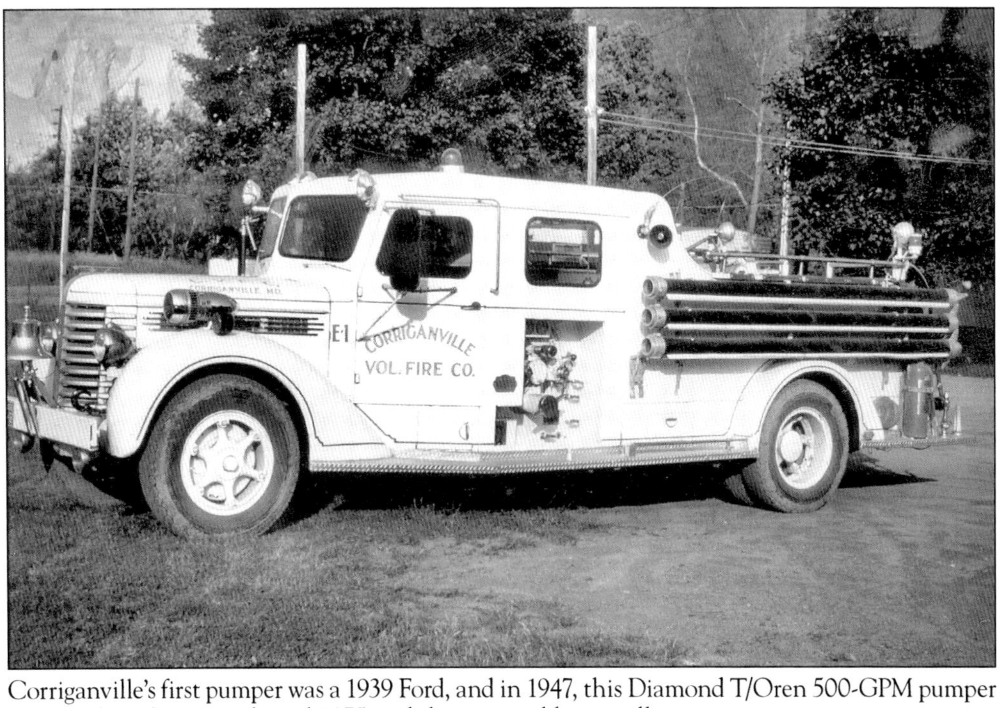

Corriganville's first pumper was a 1939 Ford, and in 1947, this Diamond T/Oren 500-GPM pumper was purchased. It served until 1975 and then was sold to a collector.

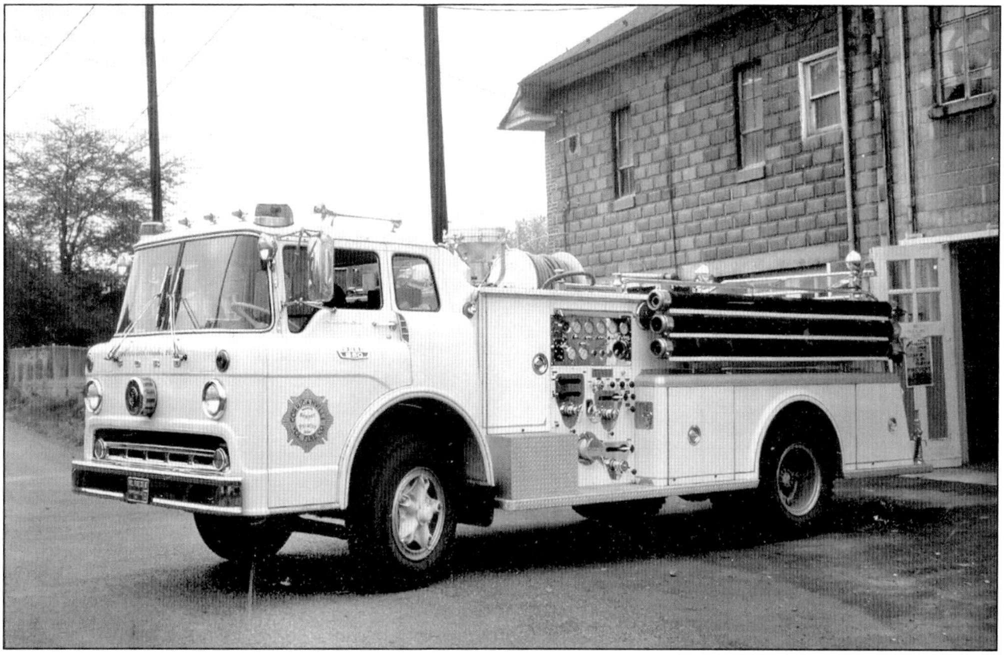

In 1965, Corriganville VFC purchased this new Ford C/Bruco pumper. It had a 750-GPM pump and carried at least 500 gallons of water. This truck probably had the first air horn in Allegany County.

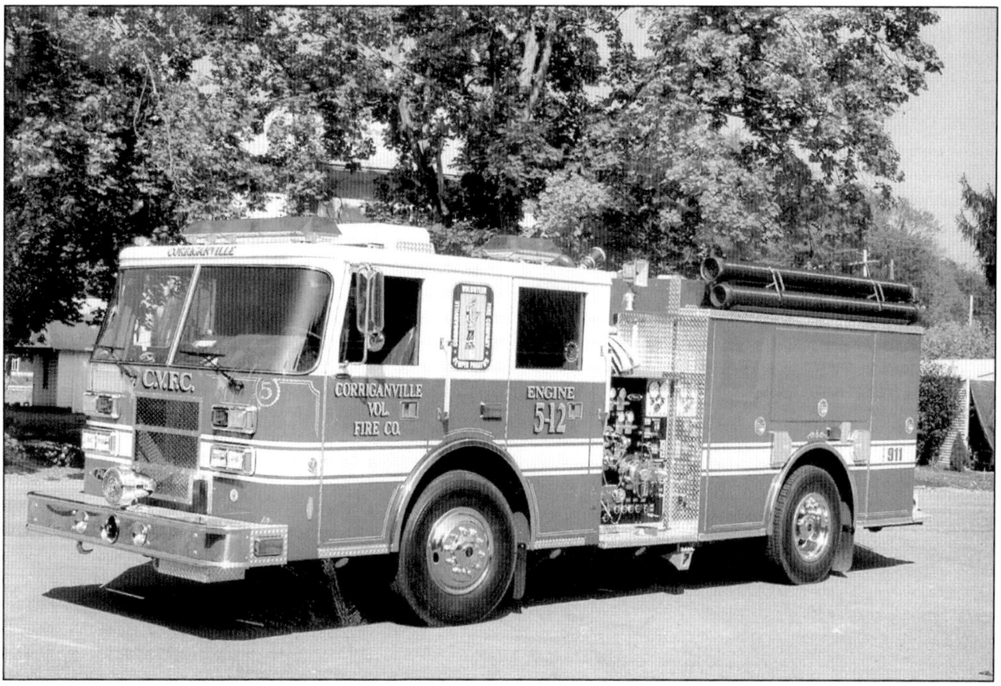

A 1987 Pierce pumper was replaced in 2000 with this Pierce Contender model pumper that has a 1,250-GPM pump and a 1,000-gallon water tank. After using a white paint scheme for the apparatus for many years, Corriganville has apparently switched to the popular white-over-red scheme.

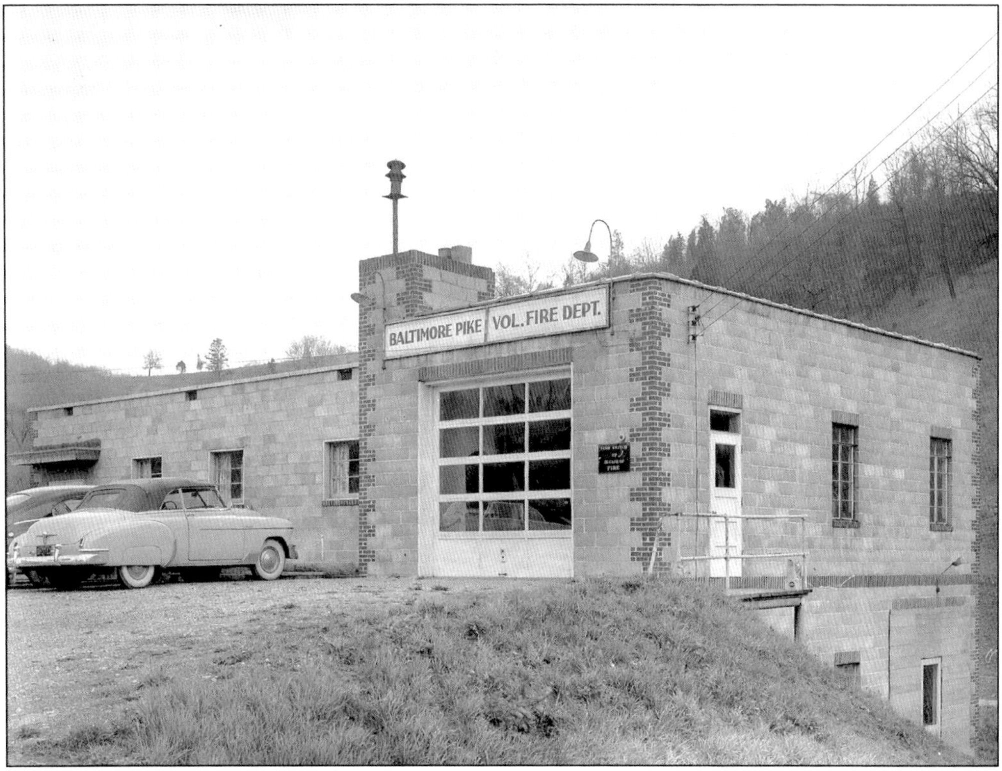

In 1948, Baltimore Pike VFC built this fire station to replace Minke's Tavern that had served as the firehouse for several years. A highway project forced the fire company to relocate and build the present station in 1967. (Courtesy Ed Bosanko, MSFA.)

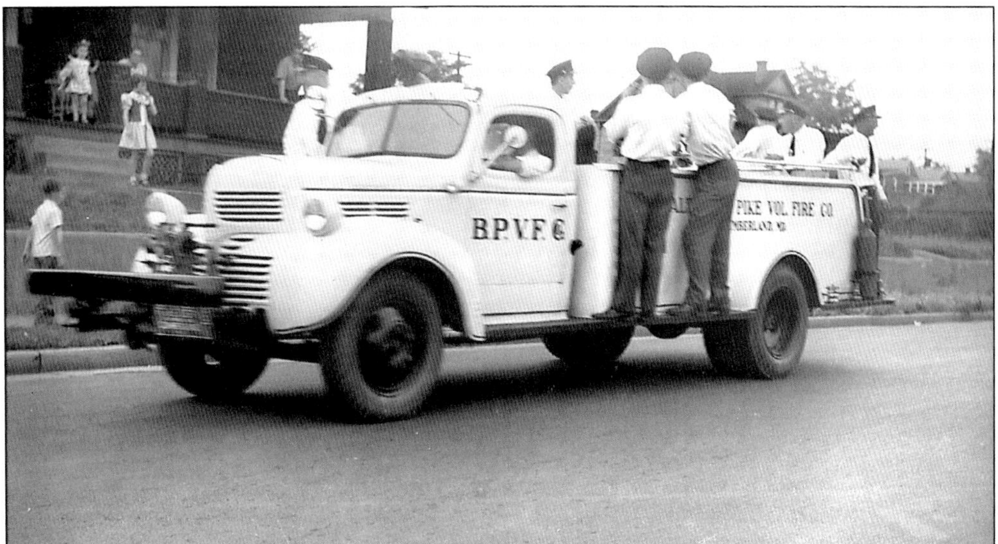

Baltimore Pike attended this parade in 1951 with their 1945 Dodge pumper truck. This truck had a small front-mounted pump and was probably fabricated by a local welding shop. (Courtesy William Snyder.)

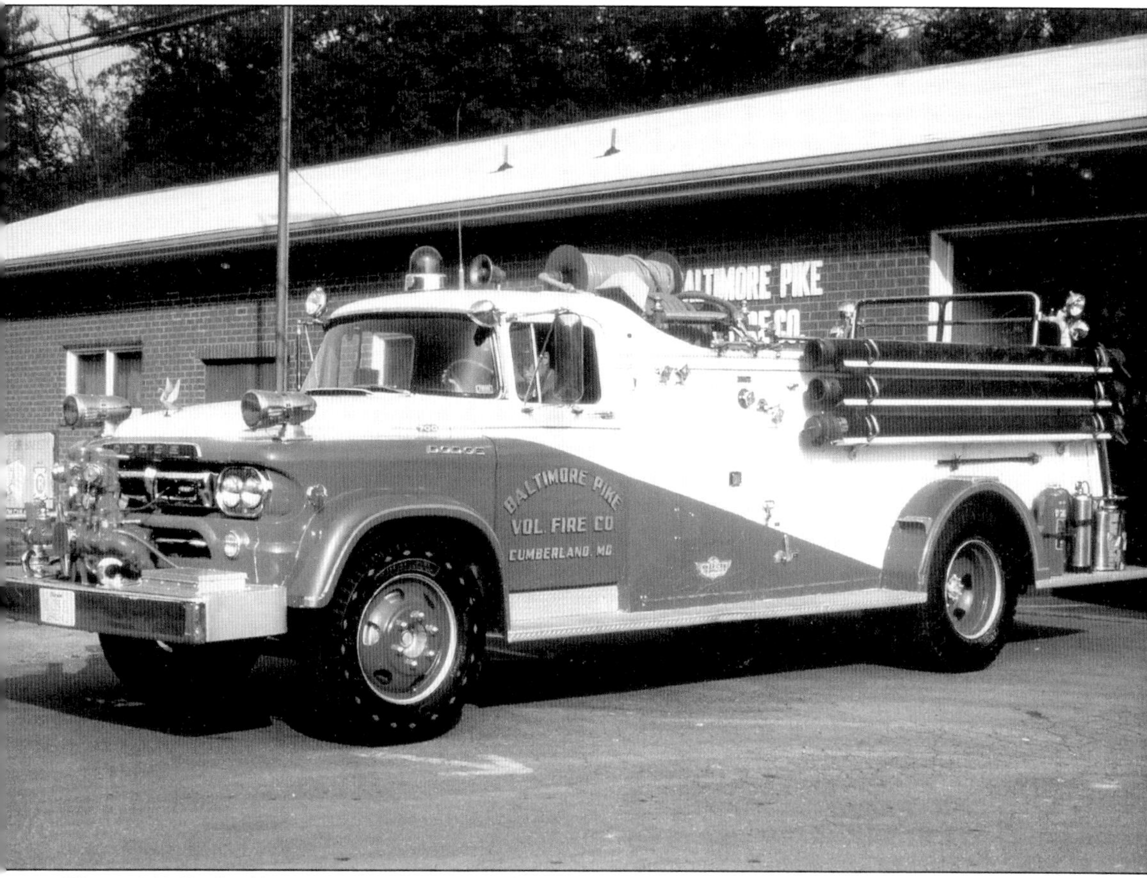

This 1960 Dodge/American pumper is a classic in both form and function. The truck had an attractive paint scheme and a 500-GPM pump, and it carried 600 gallons of water. It saw service until around 1990.

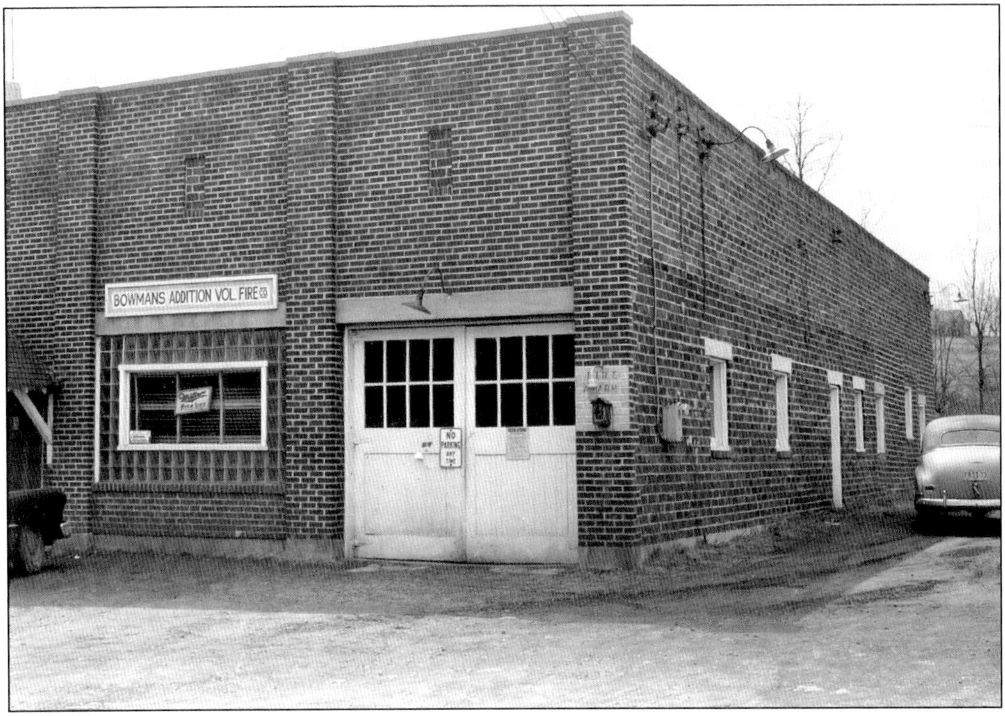

Here is Bowman's Addition VFC's station in the early 1950s. For many years, there was one engine bay and a firemen's bar attached in the social hall. (Courtesy Ed Bosanko, MSFA.)

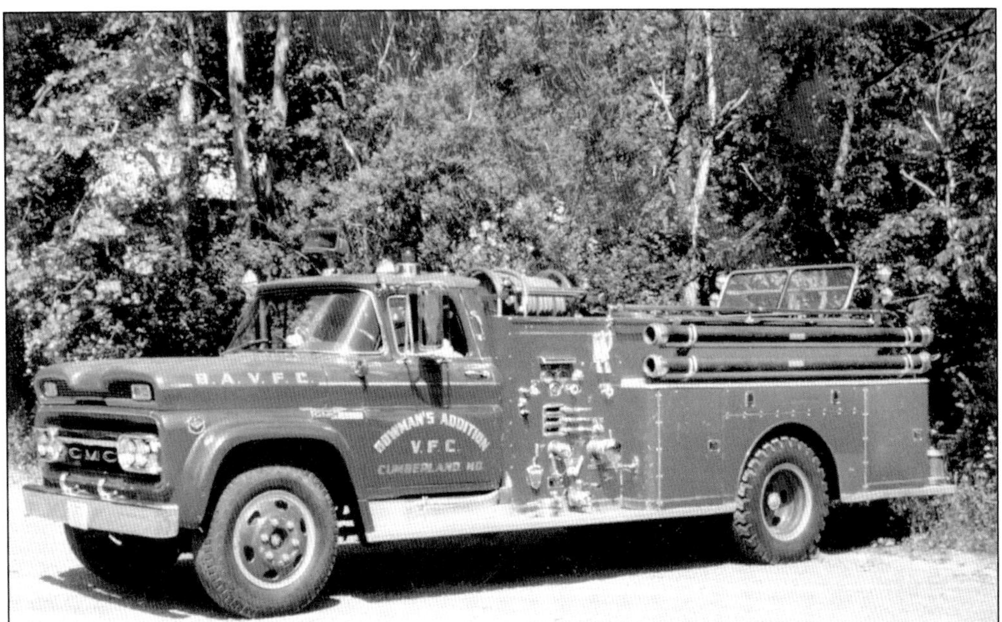

The first new fire truck for Bowman's Addition was this 1961 GMC/American pumper. It had a 500-GPM pump and carried 500 gallons of water.

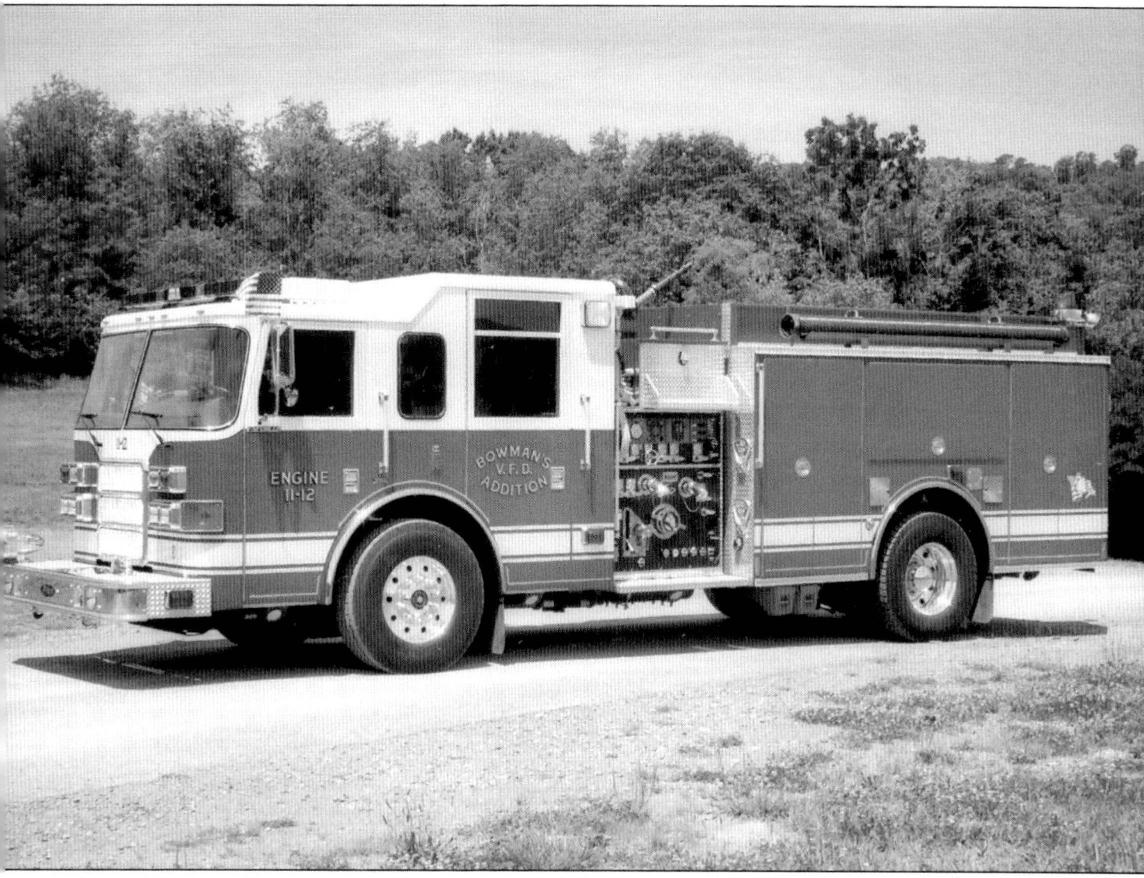

This 2003 Pierce Enforcer model pumper joined Bowman's Addition VFC and finally replaced the 1961 GMC/American pumper. It has a 1,250-GPM pump and a 1,000-gallon water tank.

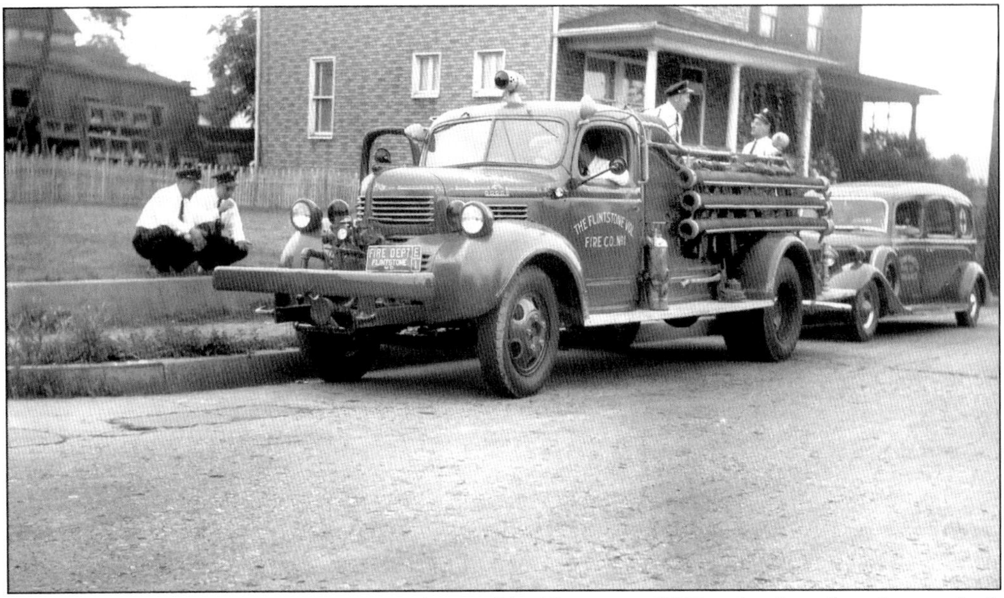

In 1945, this Dodge/Oren pumper was obtained by Flintstone VFC to be their first fire truck. It had a 500-GPM pump and carried 300 gallons of water. (Courtesy William Snyder.)

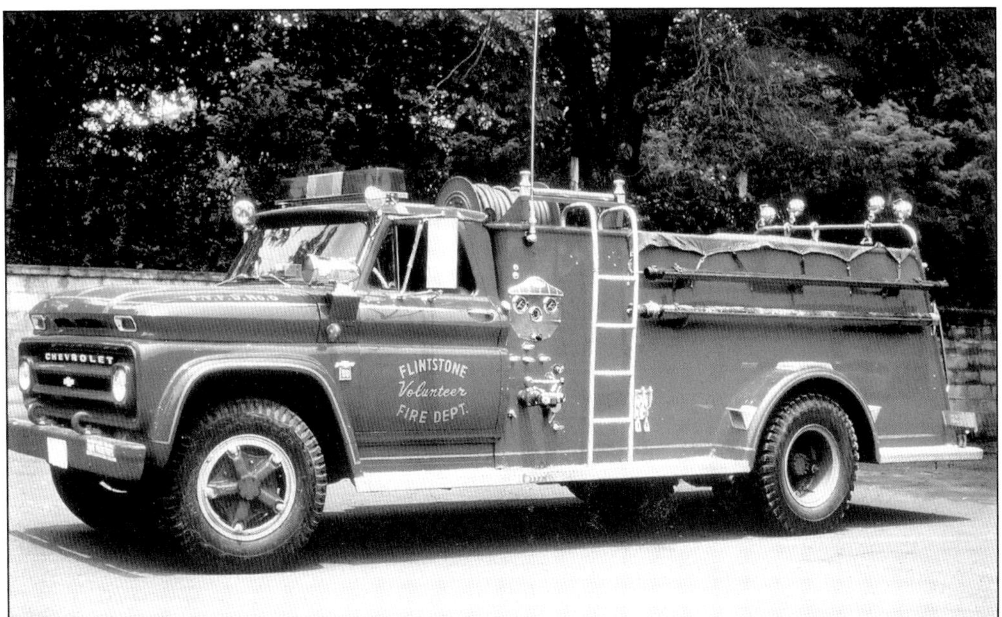

This 1964 Chevrolet Central tanker was Flintstone's first tanker truck. It had a 500-GPM pump and a 1,150-gallon tank.

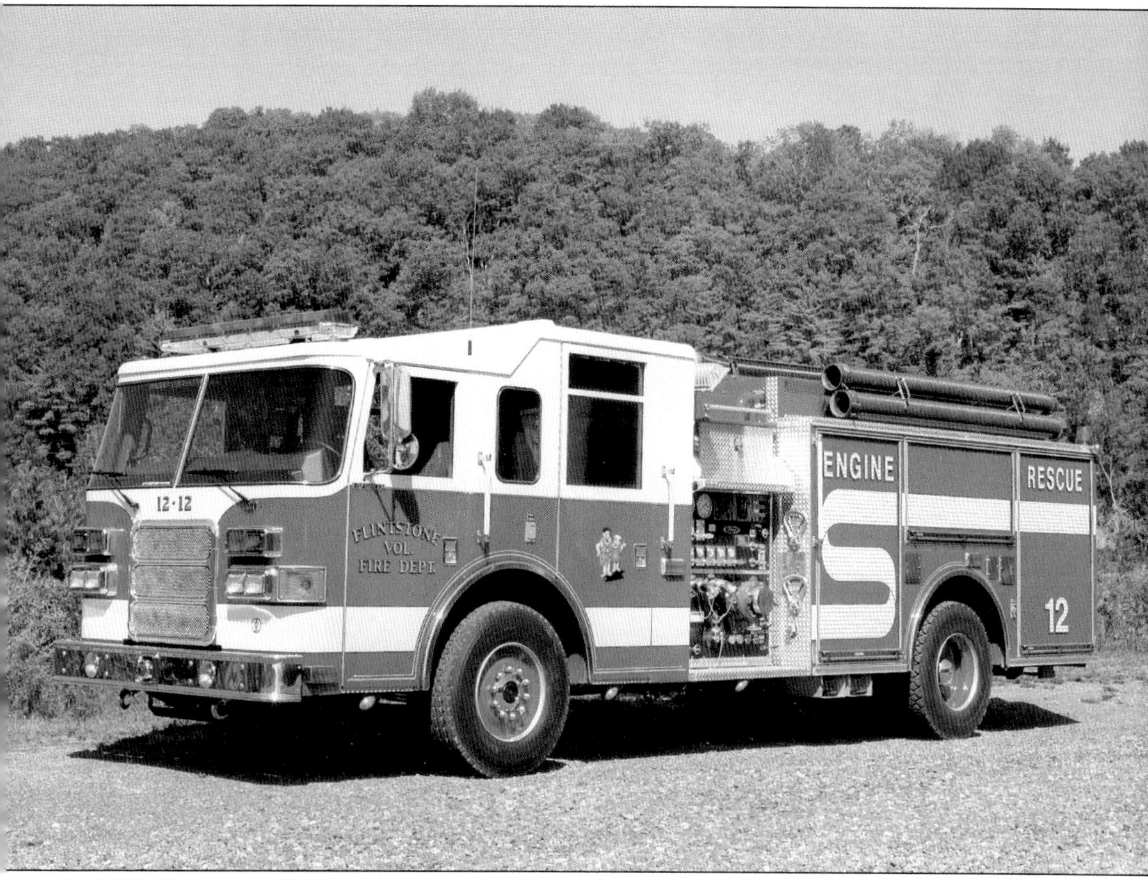

In 2001, Flintstone VFC received this new Pierce Enforcer model pumper. This apparatus has a 1,250-GPM pump and carries 750 gallons of water. It also has a full allotment of rescue equipment for auto accidents.

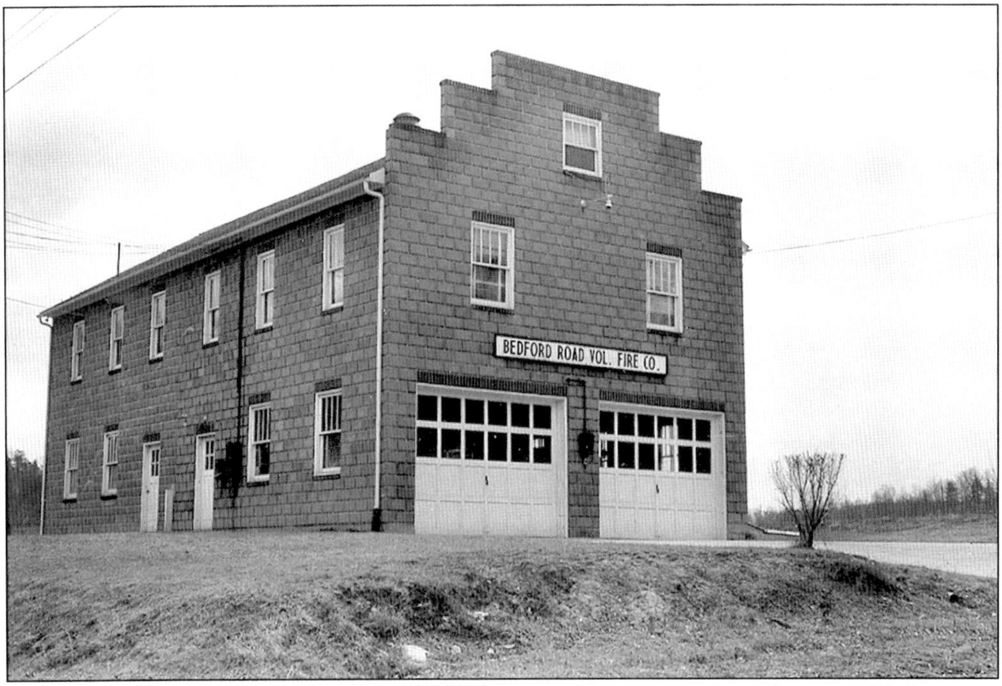

Bedford Road VFC, organized in 1943, built its original home in 1944 and it was added onto in the 1950s. This photo was probably taken after the addition was completed. In 1974 and 1984, the building was expanded to its present form. (Courtesy Ed Bosanko, MSFA.)

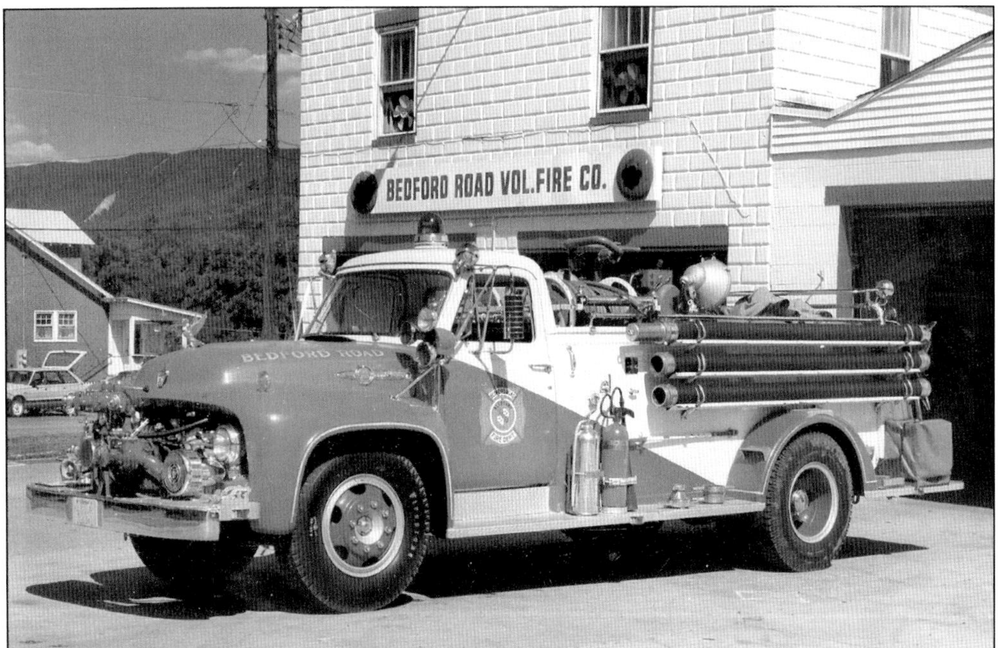

About 1958, Bedford Road VFC purchased this 1954 Ford/American 500-GPM pumper with 500 gallons of water from Forestville, Maryland. This engine served from 1958 until the late 1990s, when it was sold back to Forestville. (Courtesy Virginia McLaughlin.)

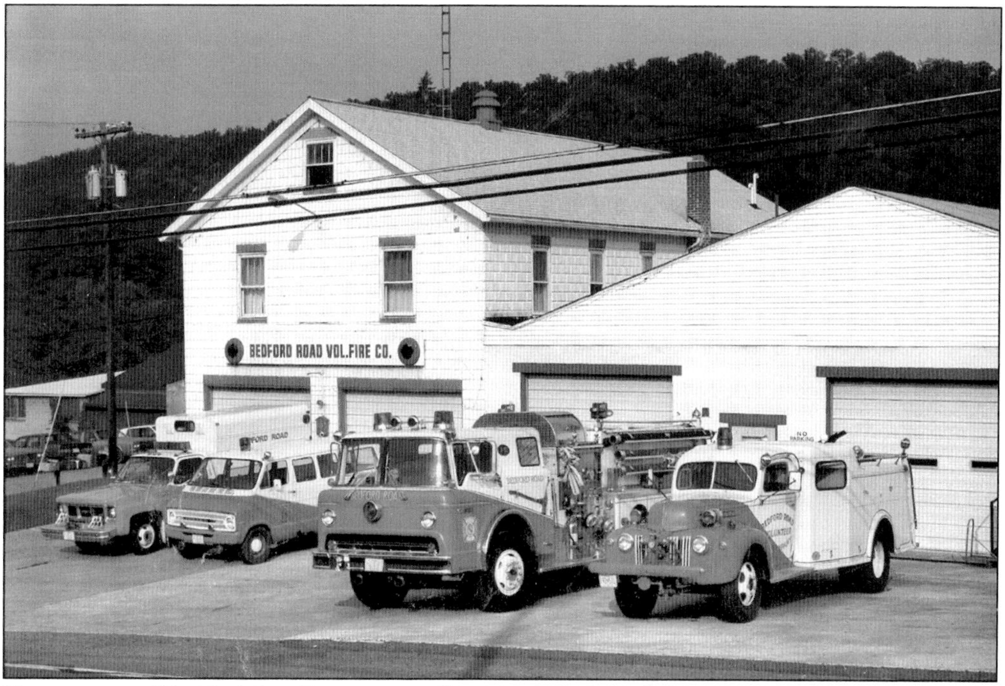

Here is part of Bedford Road's apparatus fleet in 1992. From left to right are 1974 GMC/Bruco rescue, the 1972 Dodge Van used as an air cascade truck, the 1978 Ford/Hamerly 750-GPM pumper, and the 1947 Ford/Oren 500-GPM pumper. (Courtesy John Floyd.)

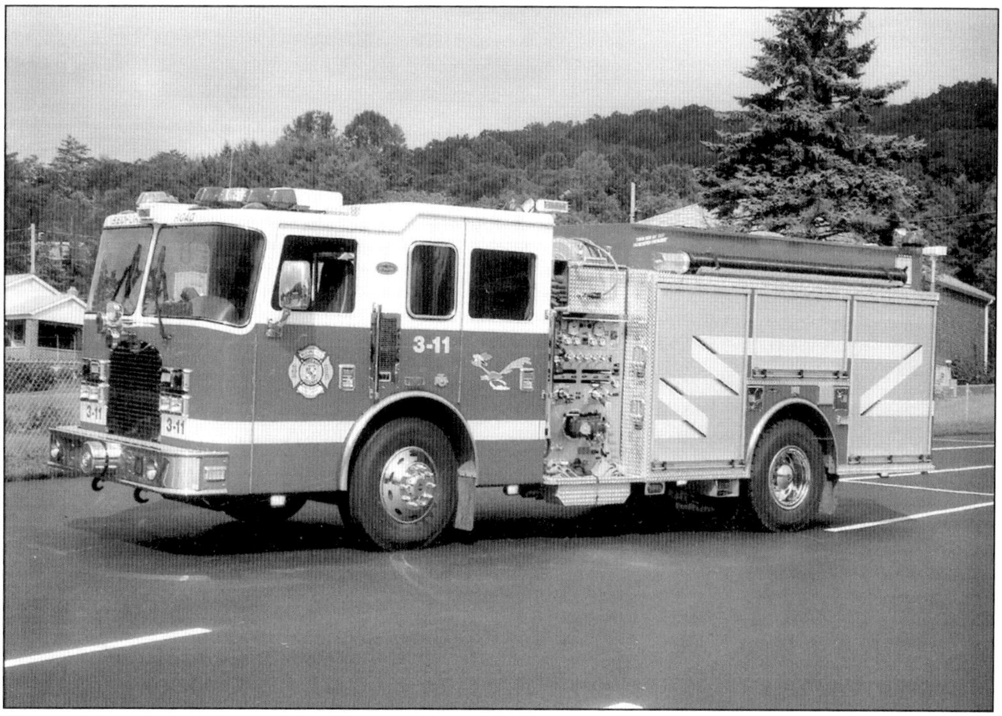

The 1978 Ford C/Hamerly pumper was replaced in 2003 by this KME pumper. This custom unit has a 1,500-GPM pump, a 750-gallon water tank, and 25 gallons of foam.

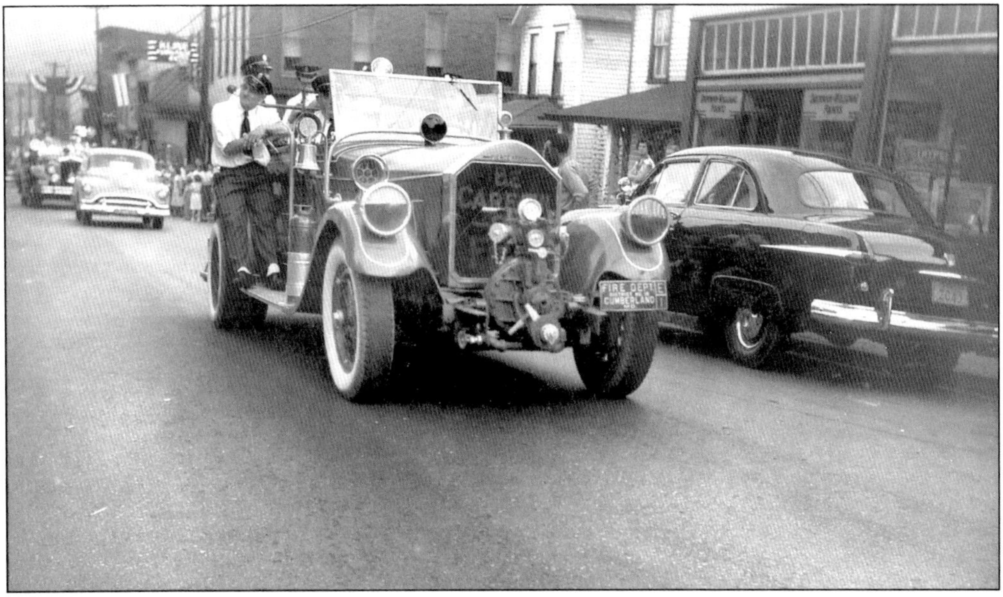

Shown here is District 16 VFC's first piece of fire apparatus. The 1929 Pierce Arrow chassis was converted into a fire truck in 1937 by a local railroad shop for the Mount Savage VFC. District 16 purchased this "truck" (originally a touring car) in November 1944 for $200, and the proud members are shown in a parade around 1950. (Courtesy William Snyder.)

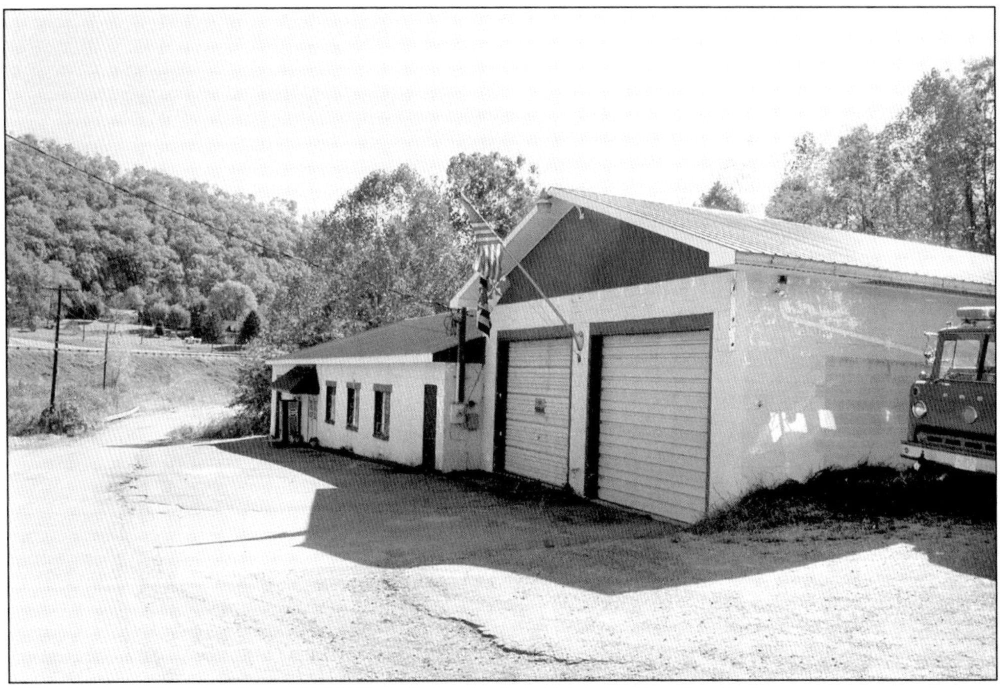

District 16 VFC's station was built in several stages over the years. This recent view shows the present form of the station. It houses four pieces of fire apparatus and an ambulance.

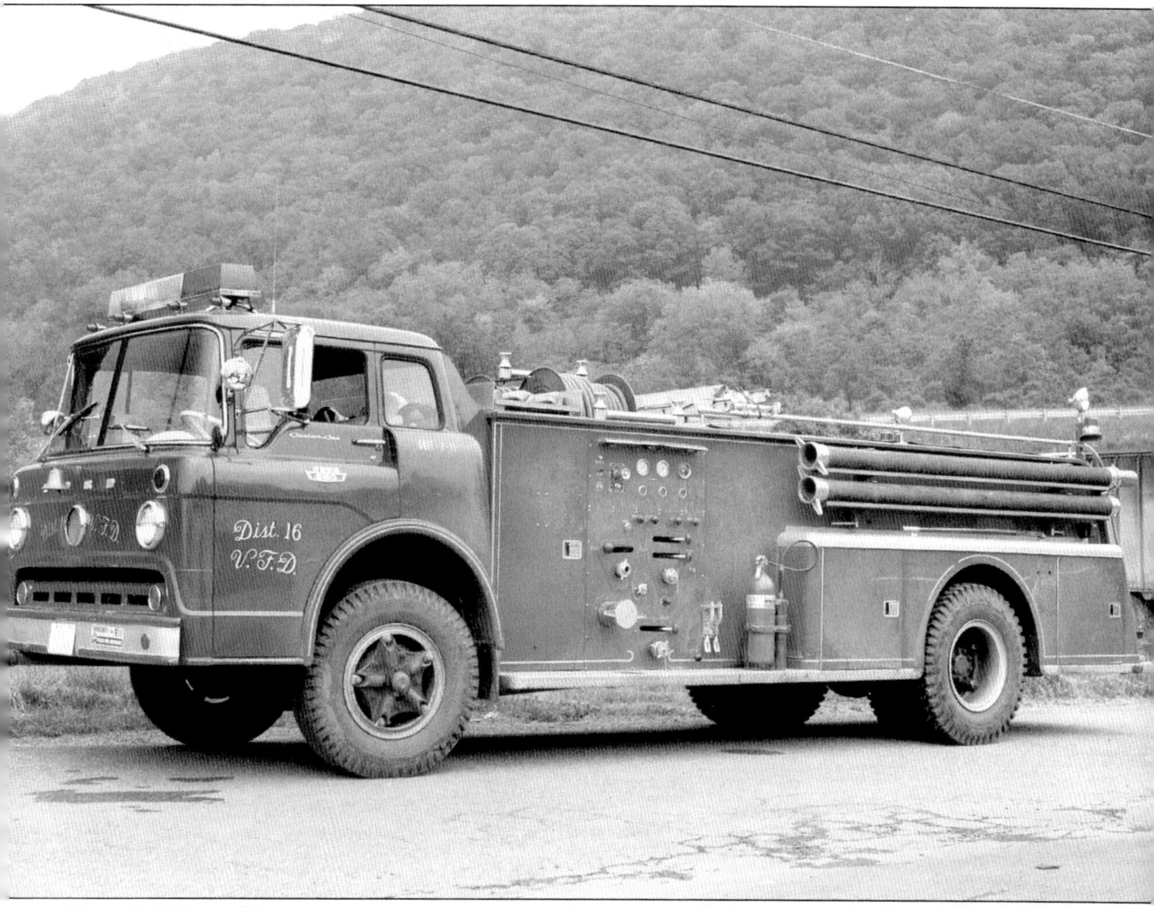

This 1965 Ford C/Bruco pumper was probably District 16's first new truck. It had a 750-GPM pump and carried 800 gallons of water. (Courtesy Virginia McLaughlin.)

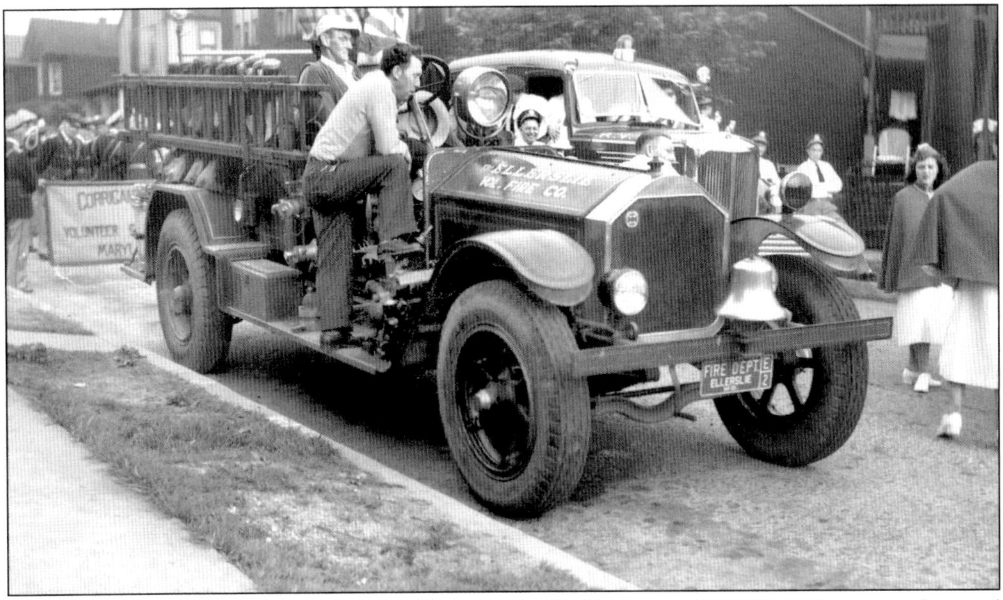

Ellerslie's first fire apparatus was this 1923 American LaFrance with a 750-GPM pump obtained from the Cumberland Fire Department in March 1947. The members were getting ready for a parade c. 1950. A 1949 Mack pumper from Ridgeley, West Virginia, is beside Ellerslie, and the marching unit of Corriganville VFC follows behind. (Courtesy William Snyder.)

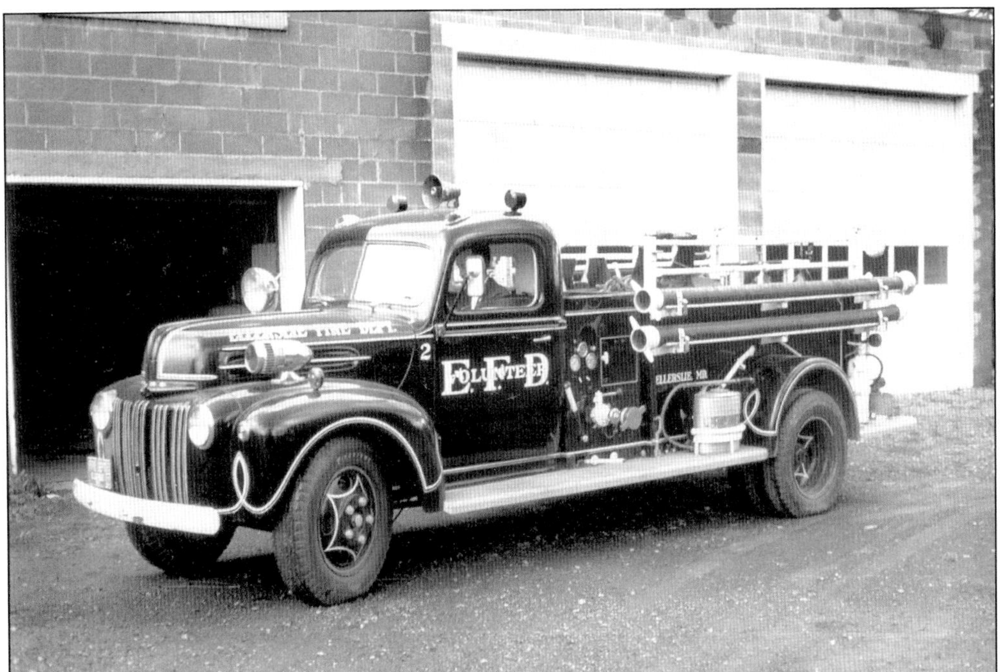

This 1942 Ford/Ward LaFrance with a 500-GPM pump was Ellerslie's second pumper. It was purchased in November 1957 from the Liberty Road (Maryland) VFC and stayed in service until 1982 when Liberty Road repurchased the truck.

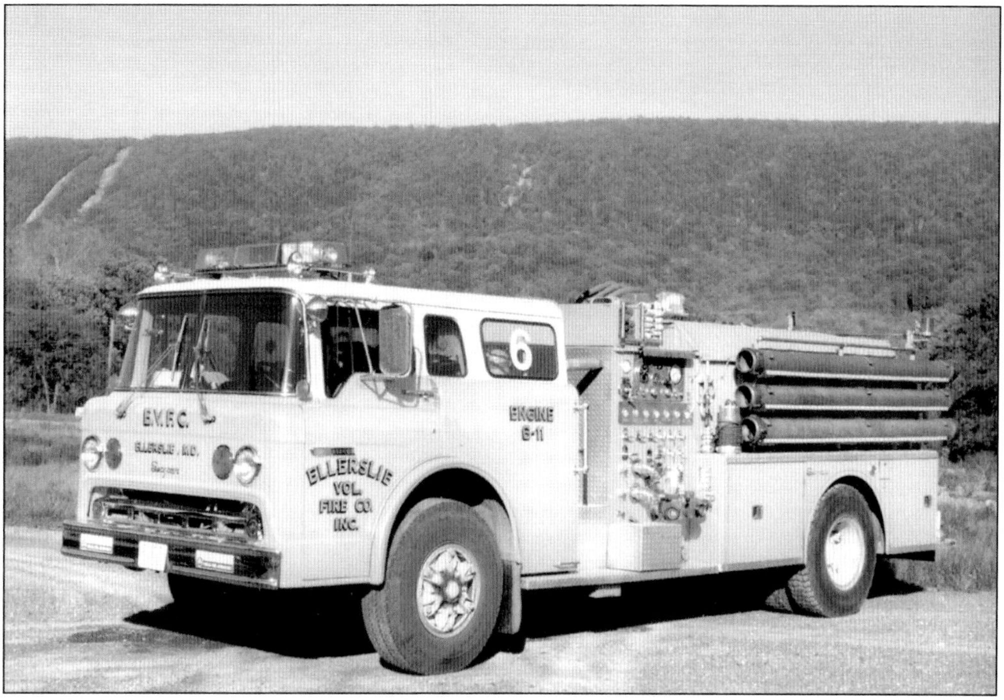

This 1975 Ford C truck was purchased and completed in 1977–1978 by Seagrave. It had a 1,000-GPM pump and carried 1,000 gallons of water. Besides being Ellerslie's first new truck, it was also the first to have the new colors of white over yellow.

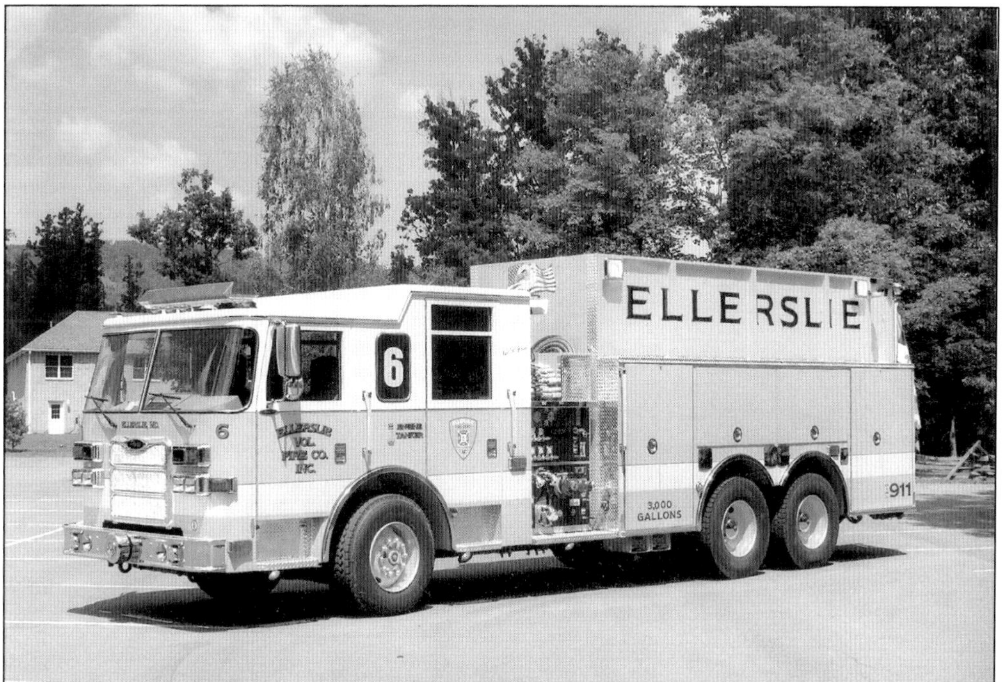

This 2004 Pierce Arrow XT model engine-tanker joined Ellerslie's fleet in late 2004. It can pump 1,250 GPM and carry 3,000 gallons of water.

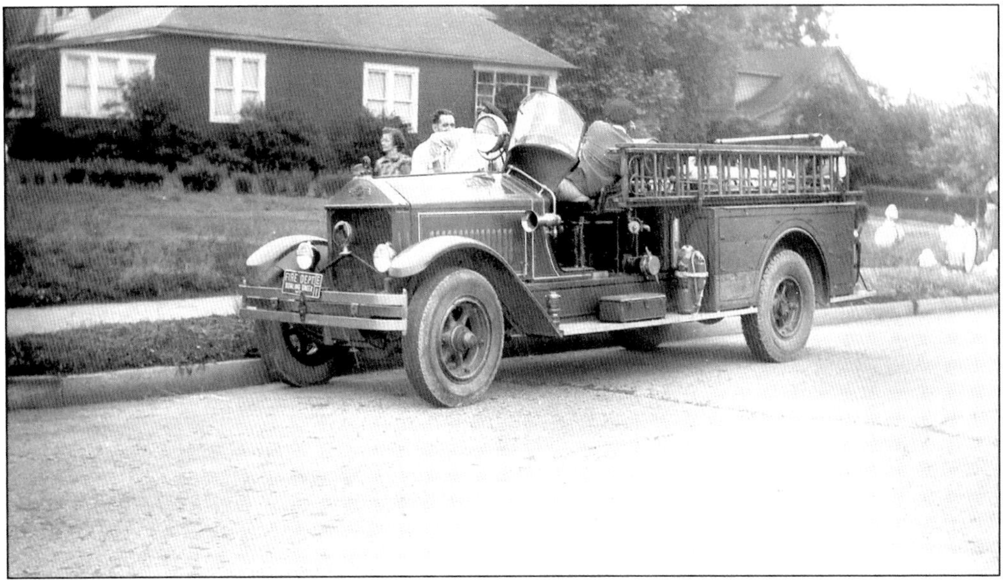

Bowling Green's first truck was this 1928 American LaFrance pumper that was bought in 1947 from Everett, Pennsylvania. It is pictured at a parade in June 1951. (Courtesy William Snyder.)

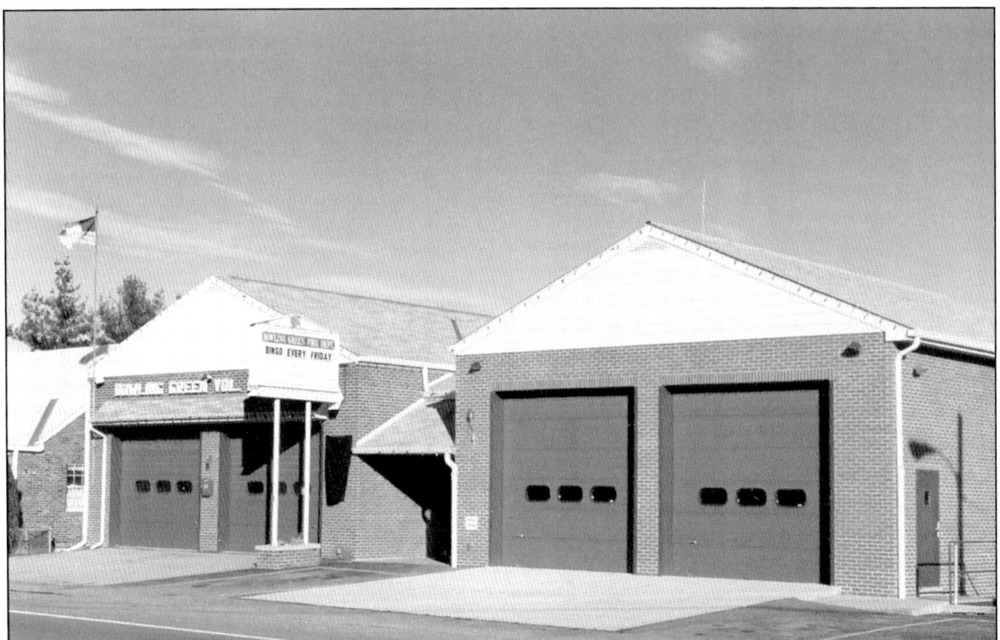

Here is Bowling Green VFC's current station in 2005. The two large bays on the right were added to the original two-bay station in 1990. The original portion was built in 1956 after the fire company's first firehouse was sold for a state highway project that was never built.

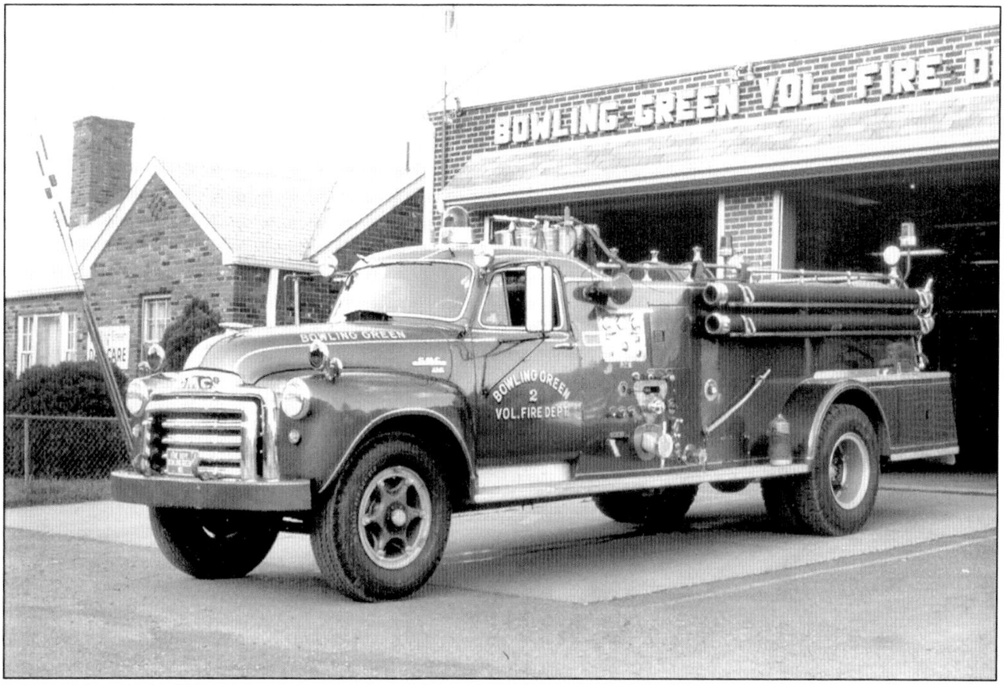

This 1954 GMC/Howe pumper was purchased and equipped with a 500-GPM pump and carried 500 gallons of water. This was the first Bowling Green apparatus painted green and was in service until 1997.

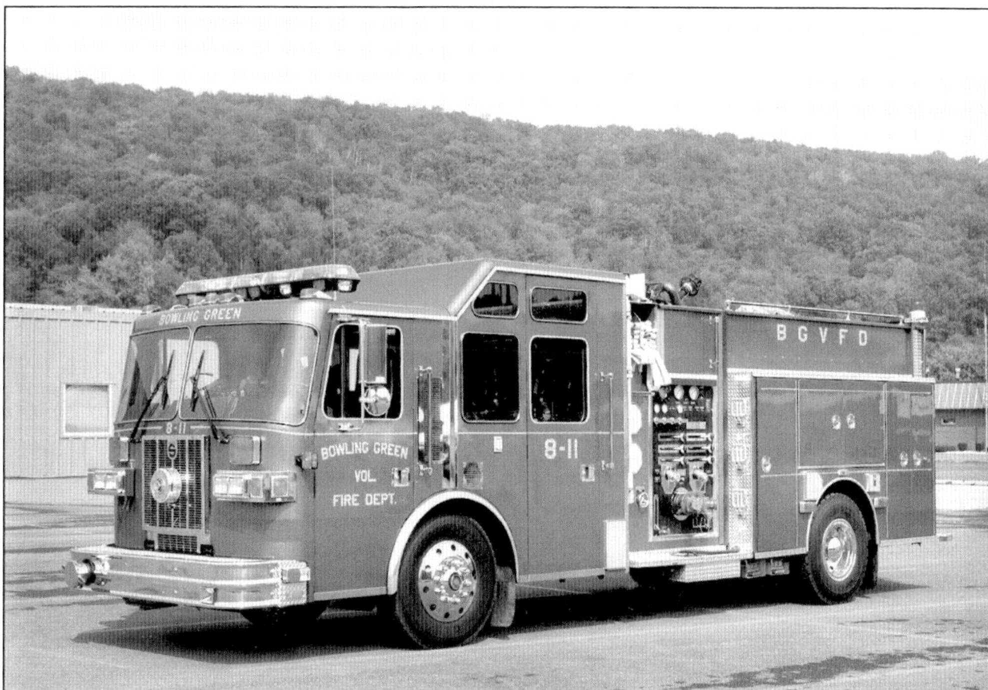

In 1997, this Sutphen pumper was delivered to Bowling Green. The 1,500-GPM unit also carries 500 gallons of water.

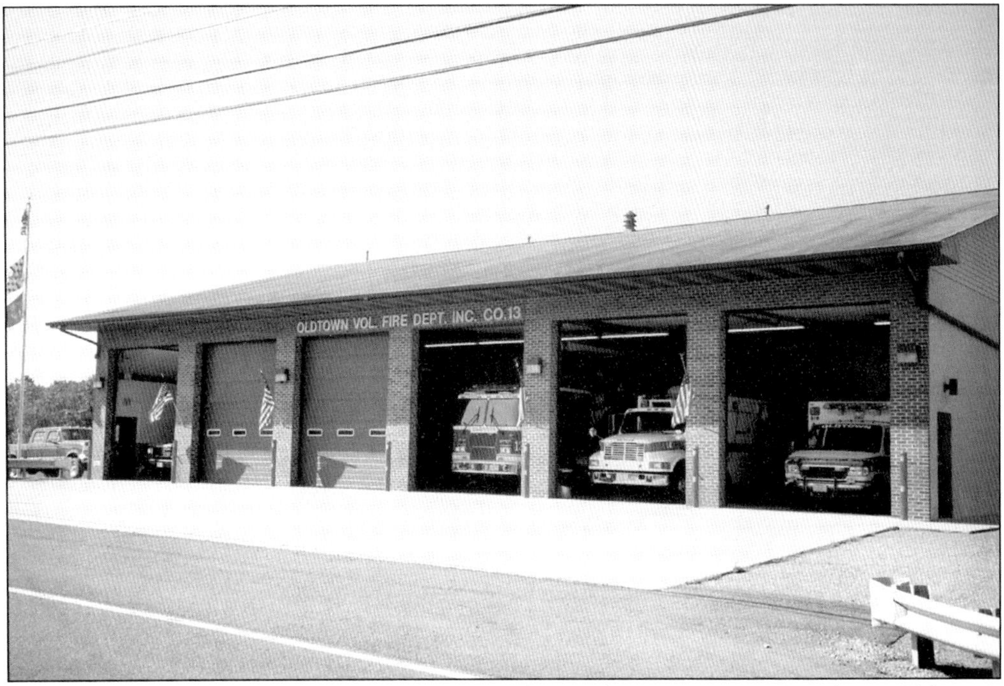

Oldtown VFC was formed in 1958 and is located in Oldtown on Route 51. This is Oldtown's current fire station, built in 1994–1995. A fire in 1994 destroyed the station erected after the company was formed.

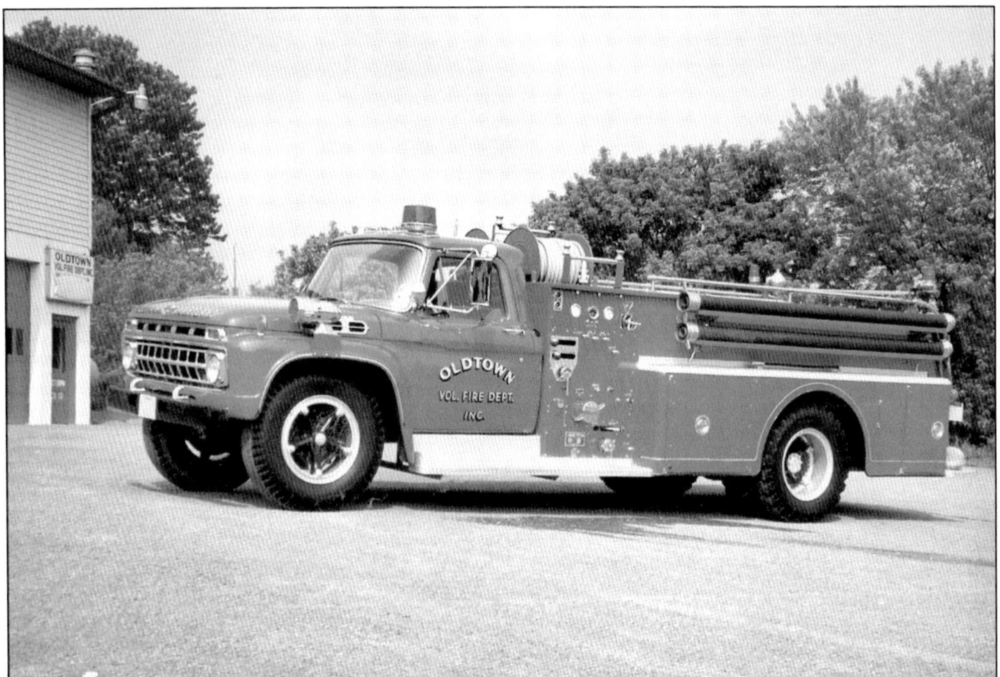

Oldtown's first new fire truck was this 1966 Ford/Hahn pumper with a 750-GPM pump and 500 gallons of water. It is still owned by Oldtown but is now obsolete for fire duty.

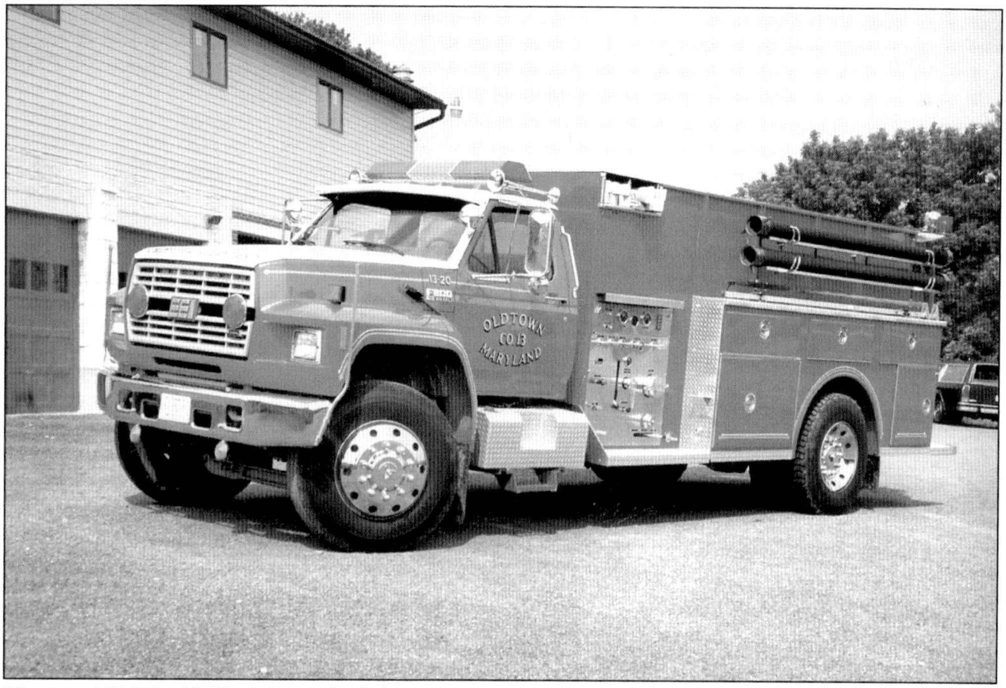

This is a 1990 Ford/EEI tanker, which has a 750-GPM pump and carries 1,500 gallons of water. EEI is a North Carolina fire truck manufacturer that made very few deliveries to Western Maryland.

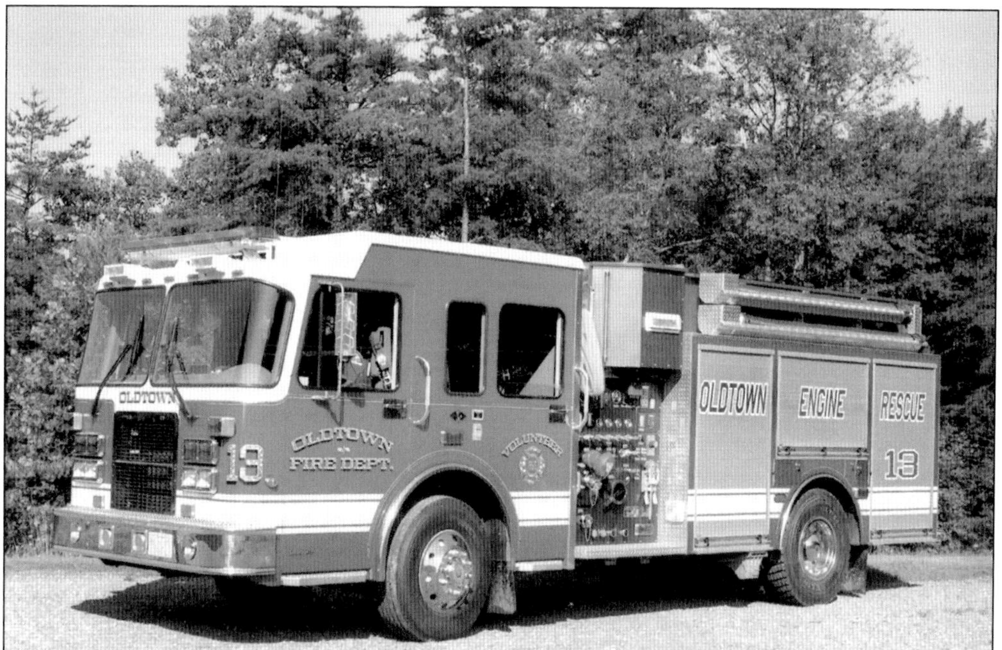

Oldtown's 2003 Spartan/4 Guys rescue engine is a versatile apparatus. Besides carrying a 1,250-GPM pump and 1,000 gallons of water, it has a full set of auto accident equipment and has a powerful light tower that can extend above the truck.

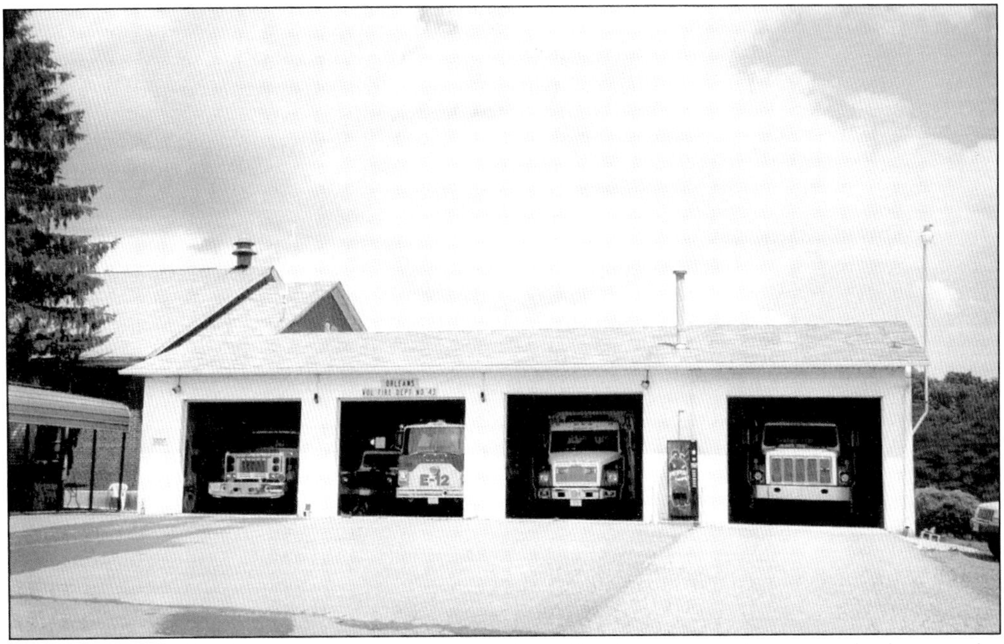

Here is Orleans VFC's station as it currently exists. The four engine bays were built onto the former local elementary school, which is still used for social events and fund-raisers. From left to right, the apparatus visible are a 1982 GMC/1997 4 Guys pumper, a 1968 Kaiser Jeep brush unit, a 1974 Mack CF/1993 Pierce pumper, a 1995 Freightliner/4 Guys rescue squad, and a 1991 International/4 Guys tanker.

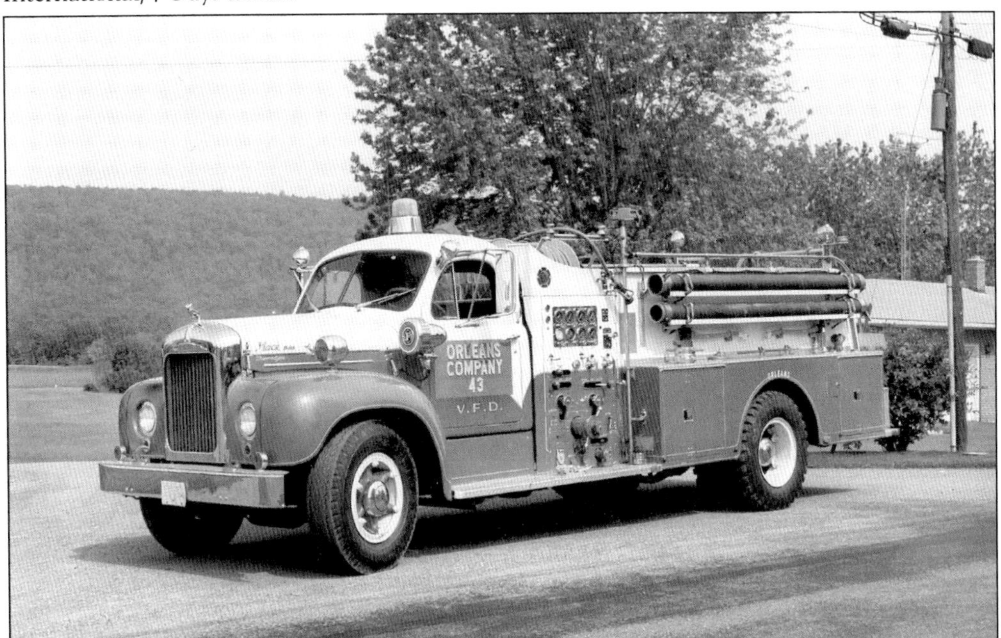

The Orleans Volunteer Fire Company purchased a used pumper in July 1981 from Elkridge, Maryland. It was a 1960 Mack B series model with a 750-GPM pump that carried 500 gallons of water. They used the apparatus until 2003. (Courtesy Virginia McLaughlin.)

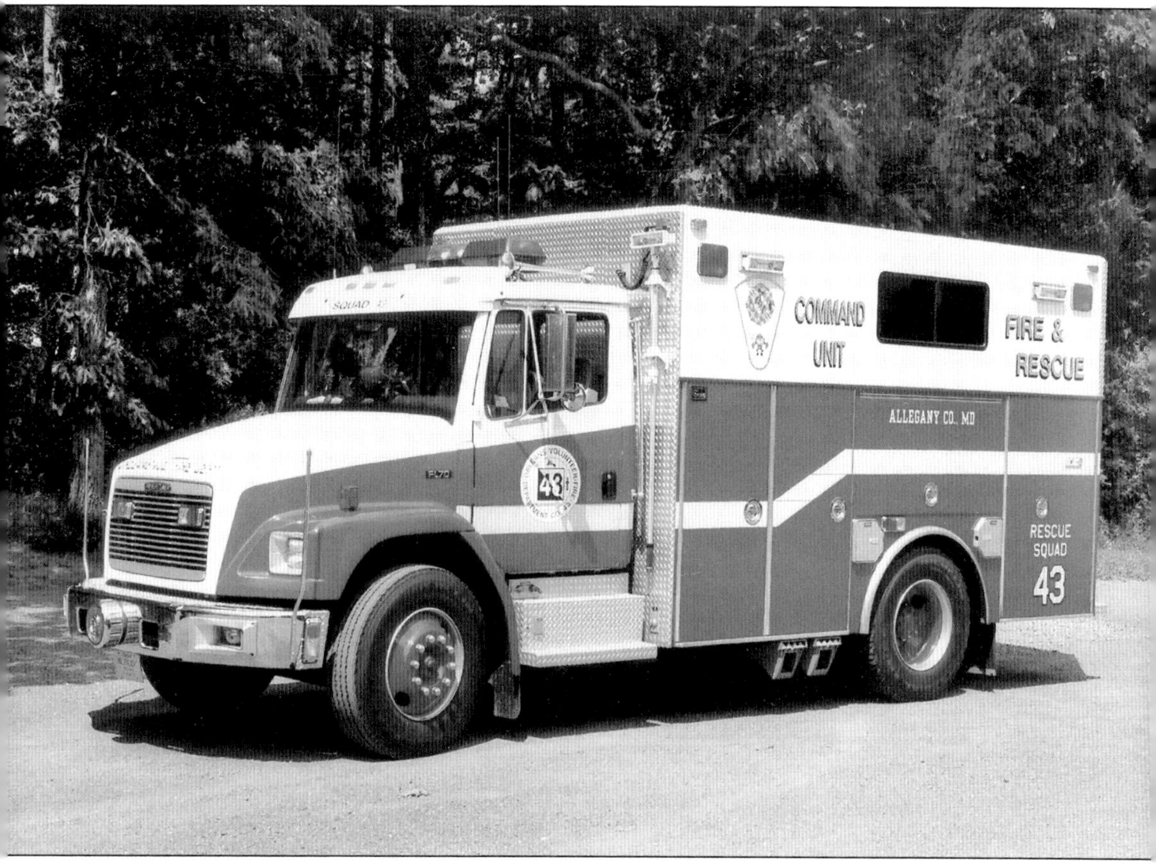

After using a jeep, step-van, and a converted ambulance as squad units, Orleans had the opportunity to purchase this almost-new 1995 Freightliner/4 Guys rescue truck in 1999 from Barton, Maryland. The rescue truck is fully equipped to handle the frequent accidents on Interstate 68.

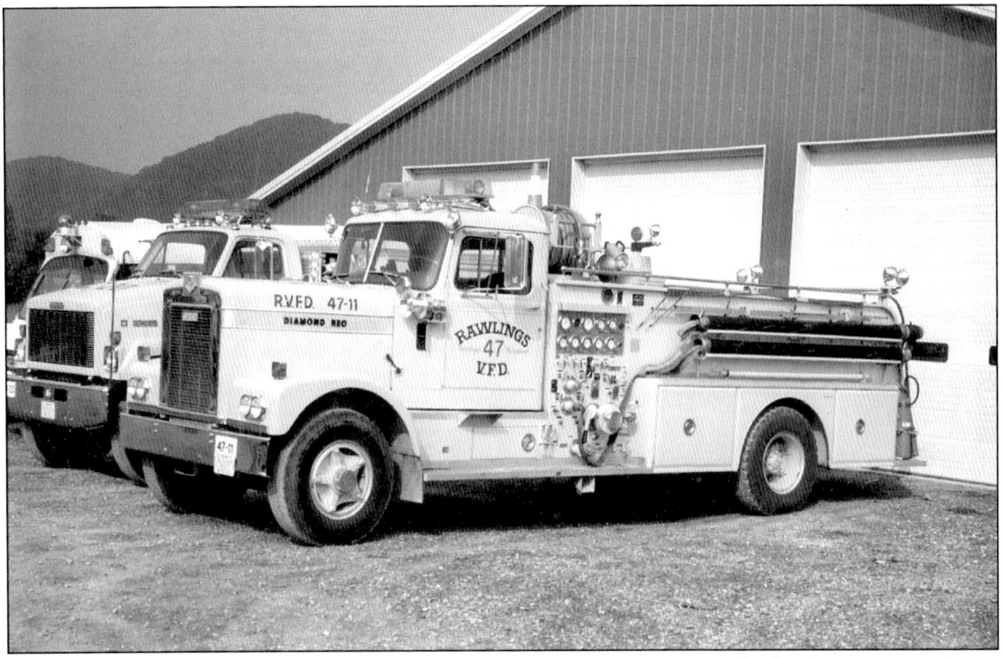

Here is the Rawling's VFD fleet of apparatus in 1992. The fire station was built after using a military Quonset hut for several years. The apparatus fleet includes, from left to right, GMC mini school bus squad, 1989 GMC/American Eagle tanker, and 1972 Diamond Reo/Oren pumper purchased used from Upper Montgomery County VFC.

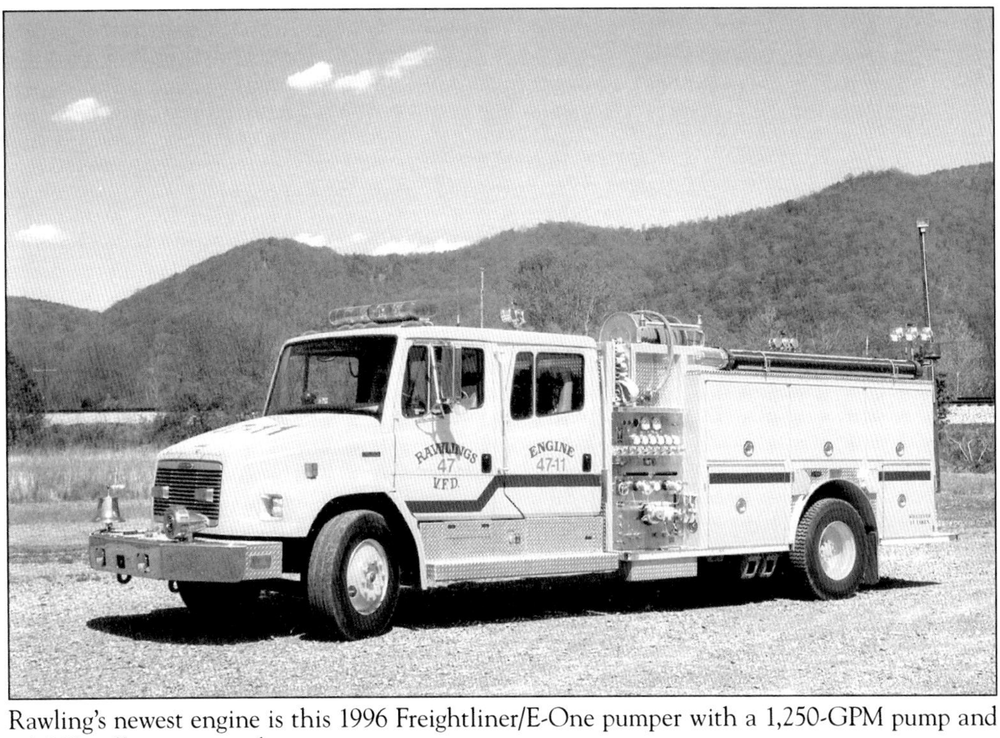

Rawling's newest engine is this 1996 Freightliner/E-One pumper with a 1,250-GPM pump and a 1,000-gallon water tank.

Five

MUTUAL AID

Mutual aid refers to the act of neighbor helping neighbor. In firefighting, timely mutual aid can be the difference between a disaster or a successful save. Mutual aid can be used for something as routine as an EMS call or for a multi-alarm structure fire. Since Allegany County began 911 service and started a central dispatch system over 30 years ago, mutual aid has been automatic to all companies dispatched by the Allegany County Emergency Operations Center. In past years, mutual aid was much more informal, and Frostburg had a separate radio system linking their surrounding companies.

Civil Defense has made a major contribution to firefighting resources in Allegany County over the years. During World War II, Civil Defense efforts were responsible for starting several fire companies. Even Cumberland had an auxiliary fire department for many years. Civil Defense continued to be active for several decades, providing an occasional rescue vehicle or surplus military truck for fire department use.

A new direction for Civil Defense in the 1970s was obtaining a surplus truck to become a county-wide air and electric special unit. This truck has been updated and continues to respond with volunteers.

The county H.I.R.T. (Hazardous Incident Response Team) was formed in 1990 from both volunteers and career staff of Allegany County fire departments.

Mutual aid can also be used with drills, particularly when burning surplus structures. However, major blazes are the main use of mutual aid, with up to two dozen companies being summoned to primary fire scenes in Allegany County. Other disasters such as auto wrecks, tornadoes, and floods have also required a large commitment of assistance, sometimes from neighboring counties in Maryland, West Virginia, and Pennsylvania.

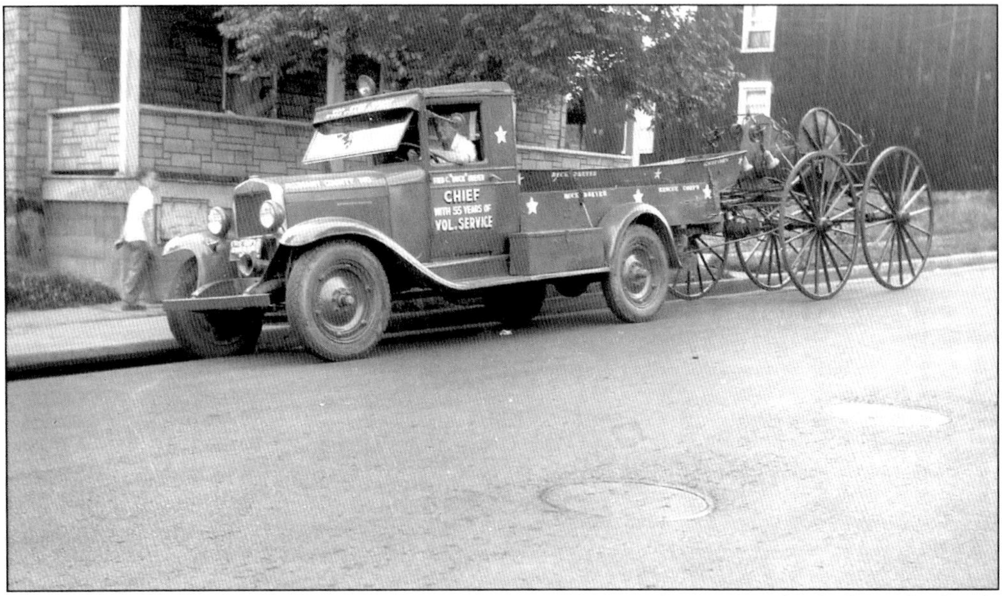

Fred C. ("Buck") Dreyer was a long-time volunteer fireman and one-time chief of Cumberland's auxiliary fire department. When this photo was taken in 1951, Dreyer operated this converted pickup with a rescue boat and towing Cumberland Hose's 1873 hose reel. (Courtesy William Snyder.)

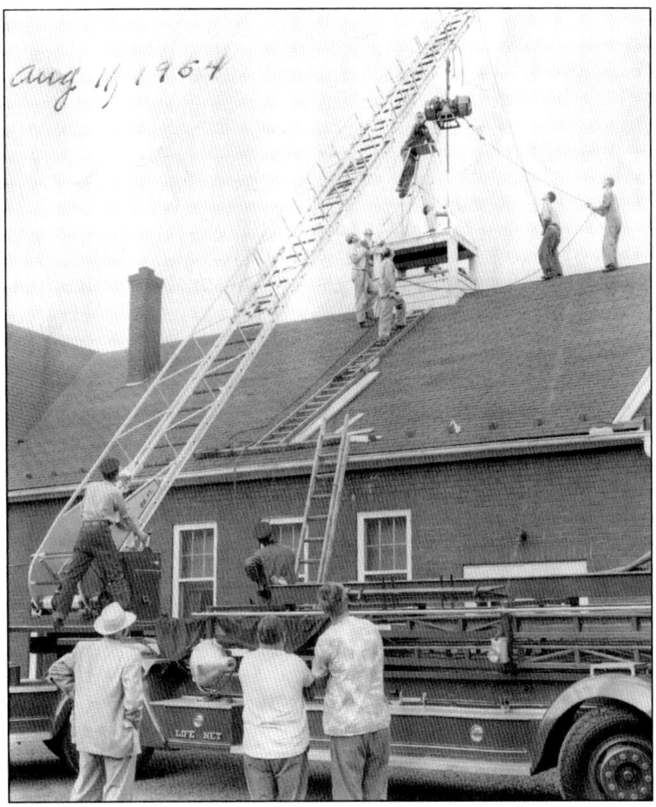

In an exercise not seen today, Cumberland Fire Department members are using the 1949 Seagrave aerial truck to install a new fire siren on top of LaVale VFC Station No. 1. (Courtesy CFD.)

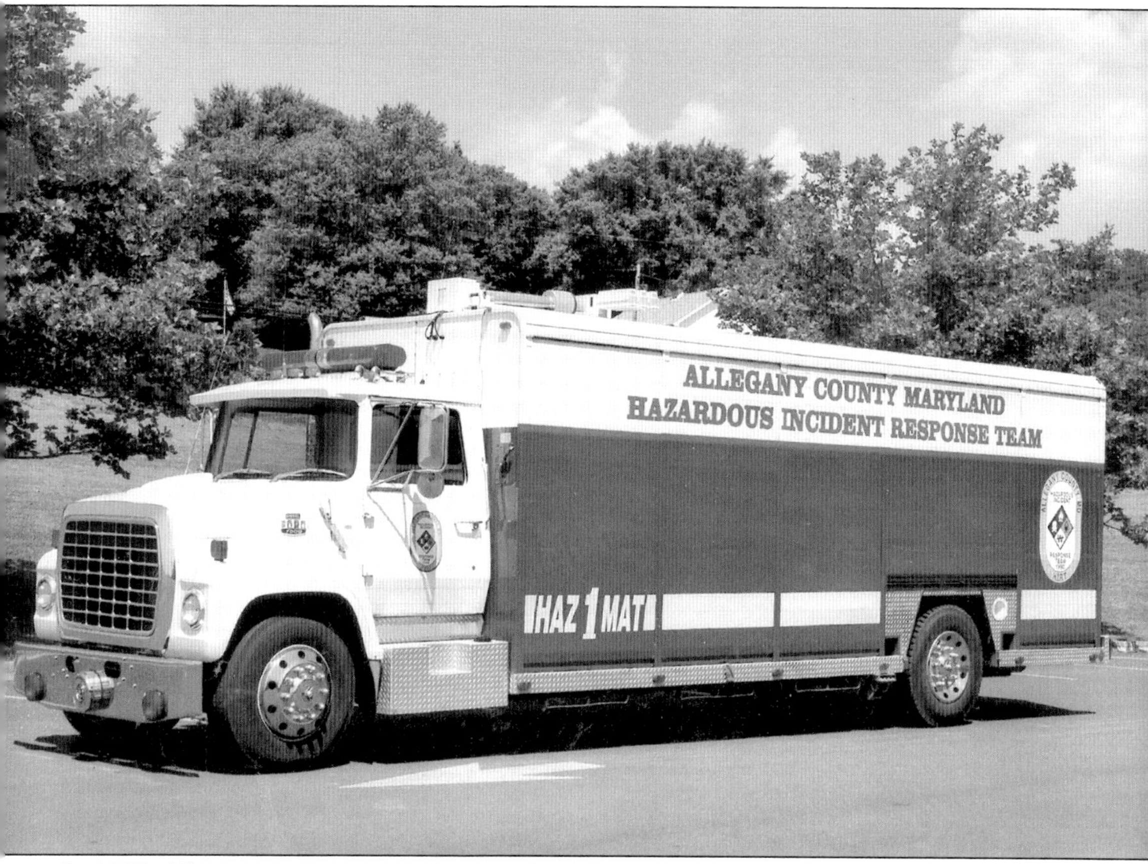

In 2005, Allegany County's Hazardous Incident Response Team responds with this converted beverage truck. The unit is a 1981 Ford/Mickey Body with modifications such as a light tower.

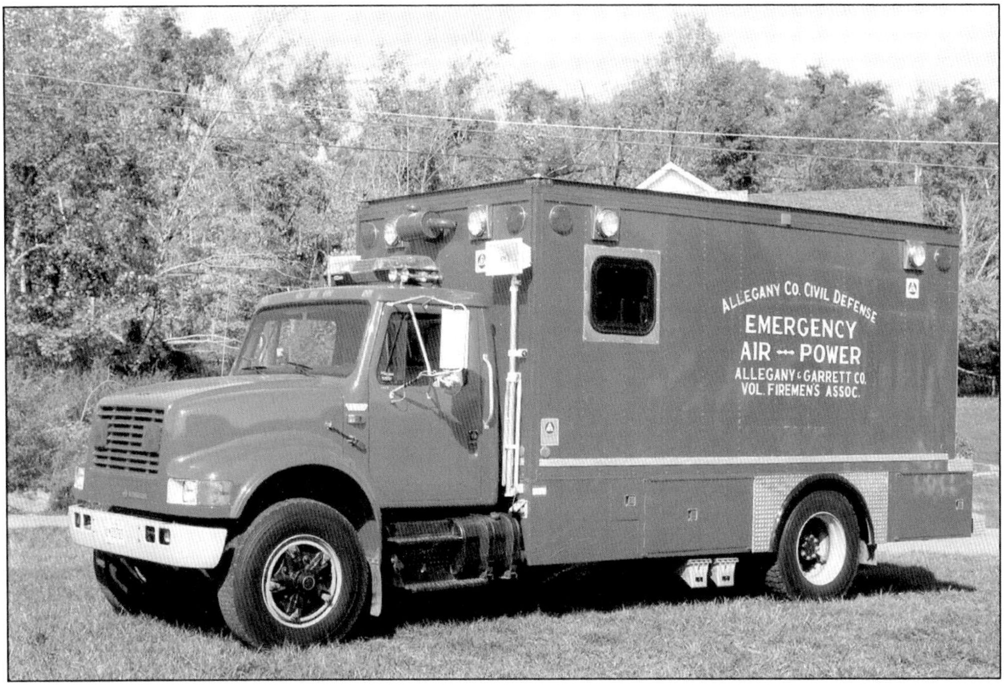

In the early 1990s, a 1989 International chassis was used to rebuild a 1971 Dodge air-cascade unit. It has been stationed at Clarysville VFC's Eckhart substation for many years.

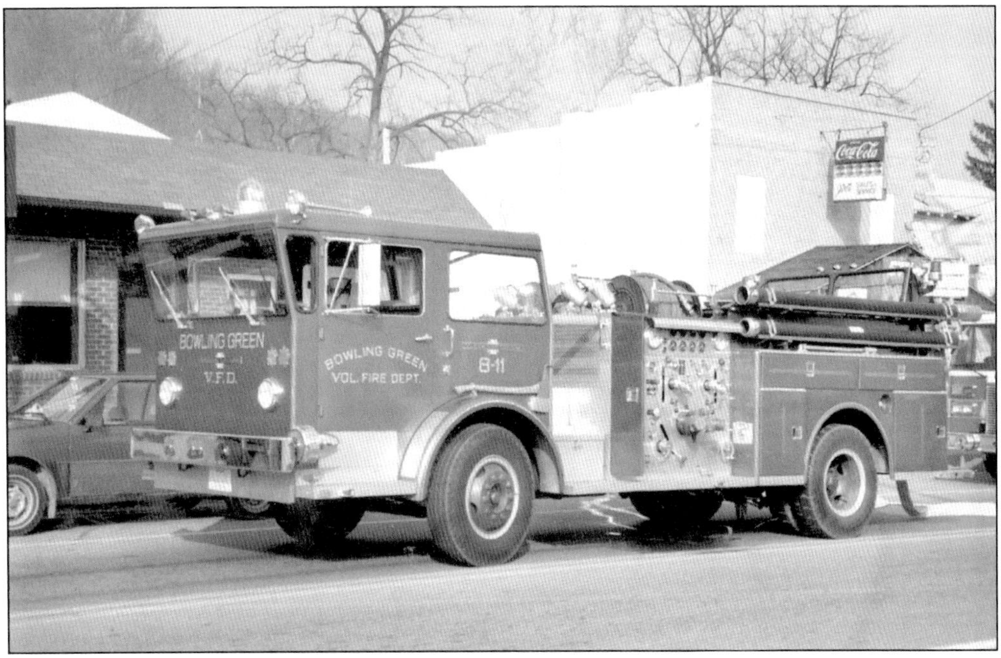

Here is Bowling Green's 1968 American LaFrance pumper in action at a structure fire in Cresaptown, Maryland, on December 27, 1991. The unit had a 750-GPM pump and carried 500 gallons of water. It responded to many of the major fires in Cumberland and nearby Allegany County for almost 30 years.

LaVale VFC's Sutphen aerial tower is seen in action at the August 20, 1996, blaze in downtown Frostburg. It is a 1982 unit, rebuilt completely in 1995 with a 1,500-GPM pump, a 300-gallon water tank, and a 90-foot tower.

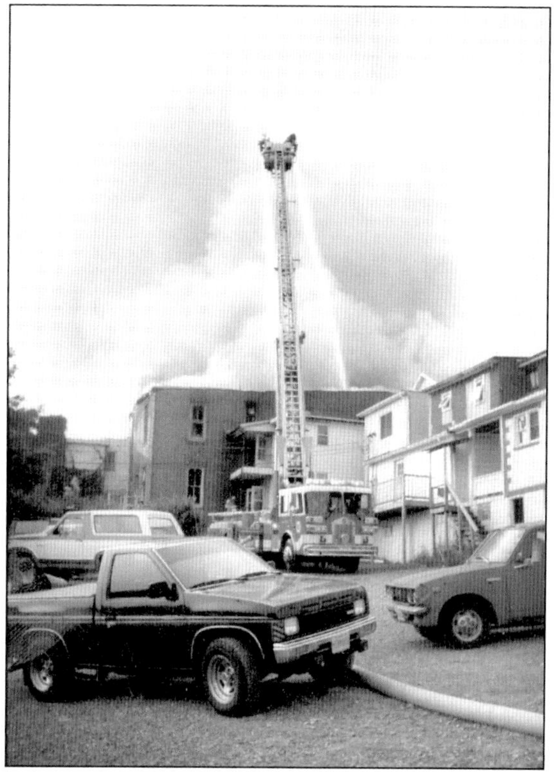

On March 10, 1999, one of the greatest losses of Allegany County's history occurred when the famous Clarysville Inn was destroyed by a fast-moving blaze. The structure, which dated from 1806, stood only a few hundred feet from the Clarysville VFC. (Courtesy Steve Bittner, *Cumberland Times-News*.)

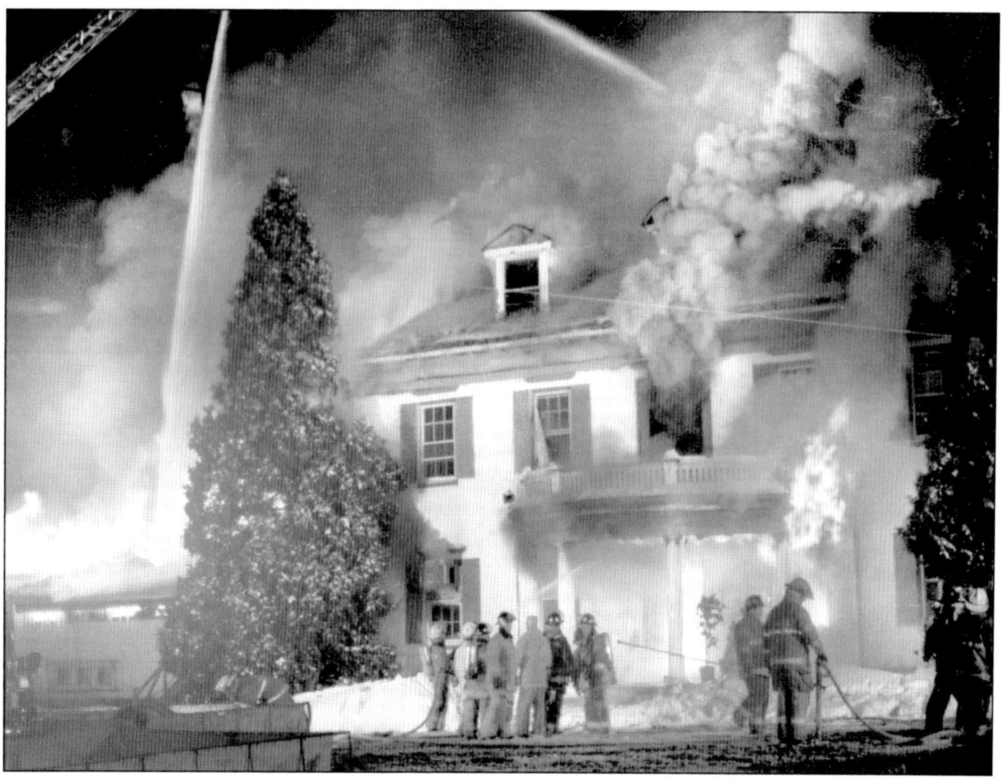

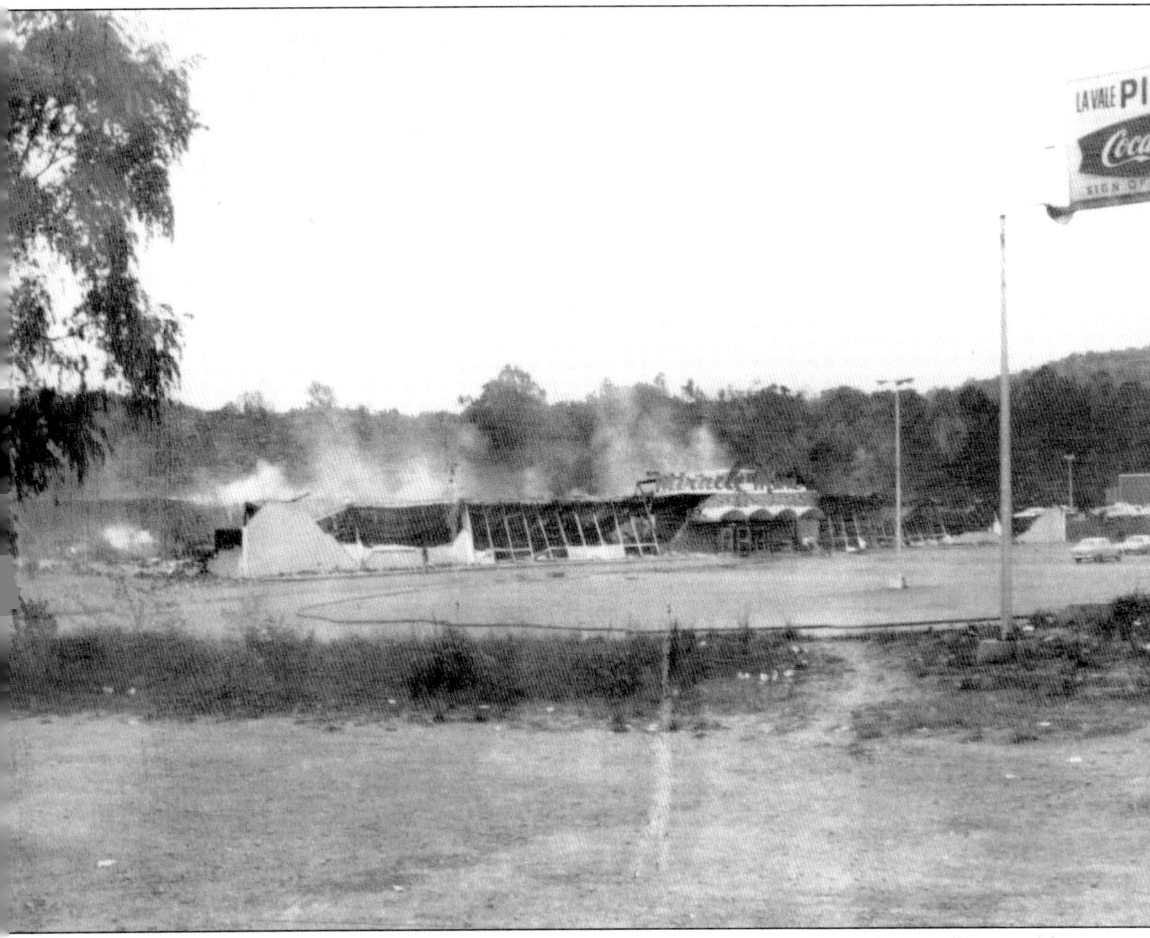

On October 16, 1966, the huge Miracle Mart Department Store in LaVale was flattened by a violent blaze fueled by exploding paint and ammunition. Water supply was the major challenge to firefighters from LaVale and at least half a dozen mutual-aid companies. (Courtesy CFD.)

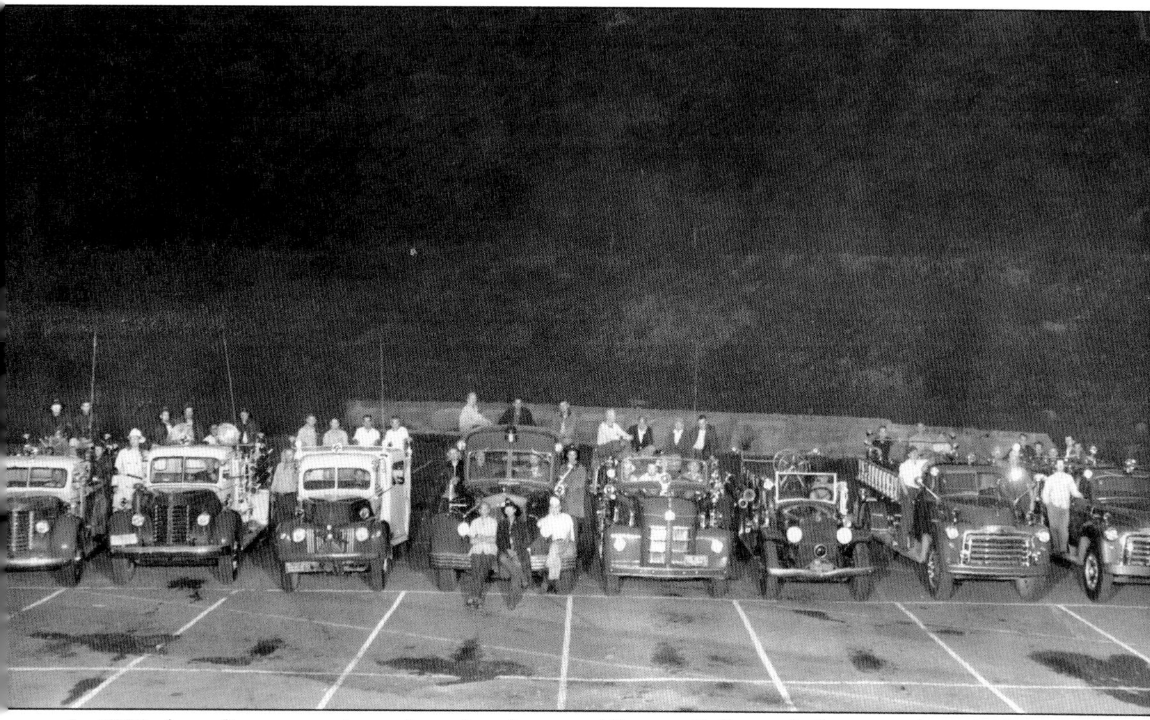

In 1955, these fire companies gathered at the new Allegany Ballistics Laboratory for a drill. The apparatus from left to right are LaVale's 1941 and 1949 Diamond T/Oren pumpers, Bedford Road's 1947 Ford/Oren pumper; Ridgeley, West Virginia's 1949 Mack pumper; Frostburg's 1949 Ahrens-Fox pumper; Cumberland's 1931/42 Ahrens-Fox pumper; Short Gap, West Virginia's 1954 GMC; and Bowling Green's 1954 GMC. (Courtesy CFD.)

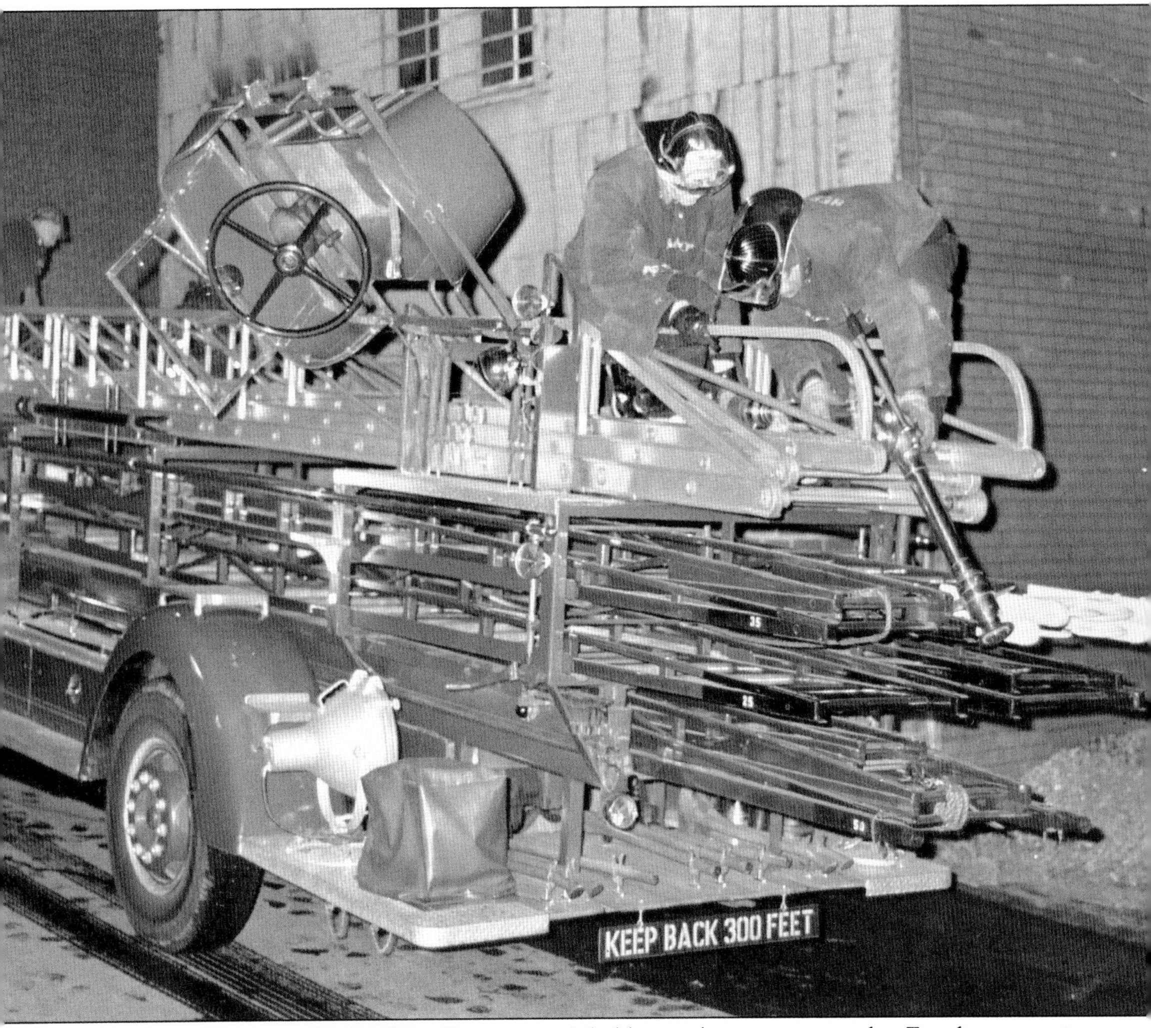

In January 1973, Cumberland Fire Department's ladder truck was summoned to Frostburg to assist with a multi-alarm structure fire. Cumberland members are, from left to right, Equipment Operator Jim Leamon, Lt. Ray McKinley, and firefighter (and future CFD chief) William Herbaugh, who is preparing to attach a ladder pipe to the aerial ladder. (Courtesy Chief William Herbaugh.)